THE A-Z OF ALLOTMENT VEGETABLES

First published in 2006 by
New Holland Publishers (UK) Ltd
London · Cape Town · Sydney · Auckland

Garfield House, 86–88 Edgware Road, London W2 2EA, United Kingdom
www.newhollandpublishers.com

80 McKenzie Street, Cape Town 8001, South Africa
Level 1, Unit 4, 14 Aquatic Drive, Frenchs Forest, NSW 2086, Australia
218 Lake Road, Northcote, Auckland, New Zealand

ISBN 1 84537 283 2

Senior Editor: Clare Sayer
Production: Hazel Kirkman
Design: Casebourne Rose Design Associates
Illustrations: Coral Mula
Editorial Direction: Rosemary Wilkinson

10 9 8 7 6 5 4 3 2 1

Reproduction by Modern Age Repro, Hong Kon
Printed and bound by Kyodo Printing, Singapore

PICTURE CREDITS

All images supplied by **Sea Spring Photos**, with the exception of the following:
Kings Seeds: cover top right, 24 bottom right, 46 bottom left, 46 bottom right, 48 bottom left, 48 bottom right.
Photos Horticultural: 23 top right, 67 top centre, 68 bottom left, 72 top right, 90 top left, 90 bottom left, 92 top centre, 95 bottom left, 96 top right, 96 centre right, 96 bottom left, 96 bottom right.
Sutton Seeds: cover top centre, 17 bottom right, 18 centre left, 18 centre right, 45 top left, 48 top right, 70 bottom right, 72 centre.

CONTENTS

A-Z OF VEGETABLES

INTRODUCTION

Seed merchants are reporting record sales in vegetable seeds. For the first time since the Second World War and the Dig for Victory campaign, sales of vegetable seeds are on the verge of overtaking those of flower seeds. Cooking and gardening are real passions of the 21st century. People want control over what they feed their families and to give them the best. Vegetables lose both flavour and goodness with incredible speed after they have been picked. Spinach loses 60 per cent of its vitamins in under three hours.

In the course of writing this book, I have enjoyed learning more about the history of vegetables and derived considerable amusement from the stories that surround them. Above all, I have developed a deep respect for the plant breeders, from the ancient civilizations through to the modern day. From a few wild plants they've bred a palette of thousands.

From the wild cabbage, kale was created, then cauliflower (or "flowery kale") – a cabbage bred for the flower buds. A cabbage grown for the young flowering shoots, broccoli, was logically described as "sprout colli-flower" in Millers' *Garden Dictionary* of 1724. Other cabbages were developed into root vegetables, including swedes, turnips and kohlrabi.

From the bitter wild carrot – the Queen Anne's Lace of our hedgerows – came sweeter forms. From the white, yellow and purple carrots of the Middle Ages, the orange carrot was developed by 16th-century Dutch nurserymen to celebrate the House of Orange. This piece of brilliance clearly met with approval at Court as the fashion of the day was to pin the feathery foliage to their hats.

Romance comes in with the great explorers of the late 15th and 16th centuries – Columbus, Cortez, Raleigh and Drake. They introduced to Europe tender vegetables from the New World – potatoes, tomatoes and French beans. Botanists travelled with them into dangerous waters. John Tradescant the Younger, keeper of gardens for King Charles I, went to Virginia to "gather up all raritye of flowers, plants, shells &c." With his cache of plants came the runner bean. Later Sir Joseph Banks presented New Zealand spinach to Kew Gardens. It was the spinach that Captain Cook gave to his crew on the Endeavour to prevent scurvy.

In Victorian England, most seed was grown and collected in Essex, a dry part of the country. The seed houses were small. Suttons Seeds (now an international giant) was started in 1806 by John Sutton, corn factor and miller. He began his seed business on his allotment as a modest sideline.

Nowadays, the seed trade is international. The giant firms commission seed growing or buy in seed from all over the world, from America and Holland to the Far East. The choice comes in a kaleidoscopic range of colours and shapes. In contrast there are also small specialist firms. We can get historic seed, not on the register, by joining heritage libraries. In response to the organic movement, there have been some remarkable breakthroughs in breeding. We now have the potato with nearly 100 per cent resistance to blight – the dreadful disease that caused the Irish potato famine in 1846. Cabbages, immune to club root, are about to be launched.

Oriental vegetables remain something of a mystery to the West. They were hidden behind the Bamboo Curtain until the 1970s, and all we have seen so far is a slow trickle of new entries in the catalogues each year. When trying to learn more about them for this book, I kept being directed back at the same source – Joy Larkcom. Somewhat at sea, and not wishing to be a plagiarist, I wrote to her for help and she kindly invited me to spend a day with her in County Cork.

I discovered that Joy had spent ten years researching Oriental vegetables, more or less from scratch and single-handed. Having mastered some horticultural Chinese, she visited China, Japan, Taiwan, Hong Kong and some of the oriental communities in the USA before writing the ground-breaking *Oriental Vegetables – the Complete Guide for the Gardening Cook* in 1991.

Recommending varieties is always difficult. Tempted to describe the more flamboyant, eye catching or newsworthy, I decided that in honesty I should only recommend varieties with a history of excellence. For this reason most of my choices carry the RHS's Award of Garden Merit (AGM). I can vouch from my own experience that these varieties bring great results. I have also included some recommended by *Gardening Which?* (GW) following field trials, and also by the Henry Doubleday Research Association (HDRA). They combine with the Centre for Organic Seed Information (COSI) to test for disease resistance in organic seed.

Don't be too cautious in your choice of what to grow. Do try distinguished old breeds or those known to be suited to your area. Consider growing some fun varieties too. You can buy a collection of different-coloured carrots from purple, through orange, to yellow. There are striped tomatoes, or black and purple ones, white aubergines the size of an egg, luminous chards, the 'walking stick' kale, or the kale that looks like a cartoon palm tree.

Try vegetables that you won't find in the shops because they have a short shelf life – the elegant claret-, gold- or green-leaved orache, or Good King Henry, an unfussy leaf vegetable, said to have been introduced by the Romans to feed the legions.

Be experimental. Grow some seakale. It is a true British delicacy as fine as asparagus and deserves to be revived. Harvest your vegetables when tiny. If you are pruning off side shoots of artichokes with small flowers to encourage the others to grow bigger, cook the young stalks. They will be tender and delicious. Pick broad beans so small they can be eaten whole in the pod like mangetout. Such delicious morsels are for the enjoyment of the home grower alone.

CHAPTER 1

VEGETABLE DIRECTORY

*Varieties marked with an asterisk * are illustrated*

POTATOES

Solanum tuberosum

Potatoes, along with bread, formed the staple diet of the poor from the late 18th to the 20th century. The first allotments in the UK sprung up in the 1790s as a response to the plight of the farm labourers. Unemployment amongst them was high with the advent of the Industrial Revolution and the Enclosure Acts had deprived them of much of the common land where they could grow a few crops and keep livestock.

At around the same time, Louis XIV, facing an uprising by the French peasants over the price of bread, gave the royal seal of approval to the potato. The military pharmacist, Antoine Parmentier, having prospered on a diet of potatoes when a prisoner of the Hanovarians during the Seven Years' War, introduced them to France with zeal. He opened potato soup kitchens for the poor. He convinced the King who held the popular view that potatoes were 'food for pigs' (and might even cause leprosy) with a bowl of delicious potato soup – known ever since as *potage Parmentier*.

The potato grows wild in the Andes mountains. It is said that Sir Walter Raleigh presented them to Queen Elizabeth I. The story goes that royal cook discarded the tubers and boiled the leaves and that, unsurprisingly, the result was not warmly received. Be all this as it may, the first published account with an illustration in England was in Gerard's *Herball* of 1597.

As the potato is related to deadly nightshade, it was regarded with suspicion at first. Another hitch was that the early varieties imported from South America were not suited to the British climate. However by the end of the 18th century potatoes had been bred and developed to become the staple diet of working men and their families. Potato breeding reached its greatest height in the 1860s when many new varieties were exhibited at the London International Potato Shows.

During the War potatoes once again were the mainstay. The Ministry of Agriculture's approved booklet entitled *The Garden A.B.C – How to Grow Food Most Economically in Small Garden Plots and Allotments*, puts the case, "Potatoes will, of course, be the principal crop unless the Garden is quite small. Even in the latter case, in times of National emergency, one should think first of Potatoes and then of Parsnips and such like, for these can be stored."

Potatoes remain one of the most popular allotment crops perhaps due to longstanding tradition or because they are so gratifying to grow. Another advantage is that they are a good way to clear poor weedy ground. Their roots help break the soil down along with all the digging and earthing up that takes place, while their leaves block out light to weeds.

Potato types

Potatoes come as first earlies, second earlies and maincrop. The first and second earlies, eaten as new potatoes, are faster to mature and less prone to the disease and the slug damage that reaches its peak in late summer and autumn. The main crop produces bigger potatoes which store well. Salad potatoes are small and waxy. They are ideal for eating cold as they hold their shape well when sliced.

Potatoes are judged on yield, tuber shape, skin and flesh colour, discoloration after cooking (harmless but unattractive), disease and pest resistance. The amount of dry matter (the proportion of starch to water content) affects how the potato cooks. Those with low starch (or dry matter) are best for boiling as they don't disintegrate easily. Those with high dry matter are good for a crisp finish when fried or roasted as they don't absorb as much fat. High dry matter also means more flavour.

Cooking

Potatoes can be boiled, mashed, baked, steamed, turned into chips or crisps, roasted or sautéed and are the base of many soups. The French make luxurious dishes of duchesse or croquette potatoes or *pommes dauphine* and *boulangère*. The Italians transform them into *gnocchi*. Potatoes are the base of the Spanish tortilla or omelette. They are a good source of vitamin C, B6 and A, potassium and magnesium, and contain complex carbohydrates for slow-release energy.

Seed potatoes

Buy top-quality seed potatoes from a reputable merchant. They come with a passport in the form of labels with the EU grades stamped on them. They will have started as certified disease-free and will carry one of the three grades – EEC1, EEC2 or EEC3. It is reckoned that the best seed potatoes available to the public are cool-stored Scottish EEC2.

A good extra precaution, if you are unsure about your soil, is to choose varieties that have been bred since 1970. They are more expensive, being subject to Plant Breeder's Rights, but they have more disease resistance bred into them.

Soil and situation

Potatoes need a sunny, open site. They are not frost hardy and hate cold wet conditions, so delay planting if necessary. The soil should have been well manured the autumn before. Break up all lumps before you begin and work the soil to a fine tilth. The ideal is pH5–6. Add a good general fertilizer or line the trench with comfrey leaves.

Chitting

Though not strictly essential, the traditional way to start off seed potatoes is by 'chitting'. Place them in egg boxes with the end with the most eyes, or buds, facing upwards. Keep them in a light place out of direct sunlight, at a temperature of about 18°C (64°F). Move them to a cooler place when they start to shoot. Around six weeks from starting, the shoots will be around 2.5 cm (1 in) long and the potatoes will be ready for planting. For fewer but larger potatoes, leave the top shoots and rub off the side ones.

Planting and earthing up

Plant in individual holes or in a trench, around 10 cm (4 in) deep, and add an extra 2.5 cm (1 in) of soil on top. Traditionally first and second earlies are planted on Good Friday (the first holiday since Christmas for the rural poor of the 18th and 19th centuries), though any time between March and May will do, depending on the weather. Aim for a month before the end of the frosts. If they come up too soon they can be protected with cloches or by more earthing up.

Plant first and second earlies 30 cm (12 in) apart in rows 45 cm (18 in) apart. Leave 38 cm (15 in) between maincrop tubers in rows 70 cm (28 in) apart. Adjust up or down according to the size.

Cultivation

While excessive water can bring on too much leaf growth at the expense of the tubers, potatoes need to be kept moist. A good dousing every two weeks in dry weather is recommended and when the flowers are forming.

As they grow, earth them up by drawing soil over them with a hoe to prevent light getting to the tubers. This will encourage a greater yield from the base. The ideal time is when the haulm (stalks and leaves) are about 23 cm (9 in) high. Bury them by about half and repeat about three weeks later, leaving about 15 cm (6 in) of the haulm exposed. Continue to earth up every three weeks until the leaves meet and shade the tubers. Try to keep the ridge slopes at about 45 degrees and the tops reasonably flat to help irrigation. Until the ridges are covered, roughen them up with a hoe from time to time to prevent a 'crust' forming.

An alternative is to grow them through black plastic on the no-dig system. This is a good way to combine weed clearance with growing crops. The only downside is that the plastic encourages slugs.

New potatoes timed for Christmas and winter eating are planted in midsummer in a barrel or bin to keep them out of the cold, wet soil. They need to be kept well watered and covered with straw or fleece when the frosts come.

Harvesting and storing

Earlies When the flowers open, the earlies are ready. They should be eaten soon after harvesting. Check the tubers for size. If you are satisfied, dig carefully from the outside inwards to avoid piercing the tubers. A flat-tined fork is useful for this.

Maincrop potatoes These should be lifted when the haulm has died back and gone brown – usually in September. If they are lifted too early they will have a soapy taste when cooked. Test by rubbing the tuber to see whether the skin comes off easily. If it doesn't, the potato has set and is ready to harvest. Choose a dry day and cut the leaves right off before you start – a sickle is the ideal tool for this.

If the potatoes come out wet, they need to be laid out in the sun for a couple of hours or brought under cover. Remove any damaged ones for immediate eating. Store the rest in double-thickness potato sacks or the modern equivalent. It is worth befriending the local greengrocer to get a supply of these. They are ideal for the purpose as they completely cut out the light while letting in air. Store in a cool, frost-free place for up to three months.

Problems

On allotment ground it is important to be extremely careful to avoid potato disease. The chances are that potatoes have been grown many times before in the same patch and it is more than likely that disease will be lurking in the soil. Problems that can arise are potato cyst eelworm, potato blight, wireworm, scab, potato blackleg, potato common scab, rust spot and slugs.

Rotation is vital. Always dig out any self-sown potatoes (known as 'volunteers') or any missed in the previous harvest. Another golden rule is never grow from your own stock or use potatoes from the greengrocer. It was the sharing and exchanging amongst small communities in Ireland that caused potato blight to became an epidemic and brought on the Irish famine of 1845–6.

Recommended varieties

FIRST EARLIES

**'Accent' (AGM)* – Introduced 1989. GW describes it as "a favourite early." Excellent and popular variety with high yield; medium-sized potatoes, waxy, with yellow flesh. Resistant to blight.

'Amandine' (AGM) – Unusual dual-purpose potato. When small, it is good for salads, when allowed to grow big it is excellent for baking. High-yielding and uniform. Pale yellow, waxy flesh. Oval tubers.

'Foremost' (AGM) – Introduced 1954. A favourite for flavour. Highly popular new potato. Waxy, and doesn't disintegrate or discolour when cooked.

'Red Duke of York' (AGM) – Introduced 1942 Holland. Large deep red tubers on a vigorous plant. Reliable with great taste and is one of the very few heritage varieties to be taken up by the supermarkets. An excellent all-rounder. Can be left to mature as a second early.

'Winston' (AGM) – The first baking potato of the season, good for showing. Large, even tubers. Bakes beautifully and doesn't discolour. Delicious.

SALAD POTATOES

'Annabelle' (AGM) – Very much like 'Amandine'.

'Cara' – Introduced 1976. Both early and late. Vigorous and high yielding with a mild flavour and good disease resistance. GW recommends it as "best if blight is a problem." For blight resistance GW also recommends the new varieties *'Valor'* and *'Verity'*.

'Charlotte' (AGM) – GW describes it as "a very popular second early." Uniformly oval. Waxy with a good flavour. The tubers have blight resistance, but not the foliage.

'Concorde' (AGM) – Introduced 1988. Good waxy yellow potato with high yields. Eelworm resistance.

'Pink Fir Apple' – A late main crop. A great old favourite, widely grown in Victorian kitchen gardens. It is knobbly (almost impossible to peel) with a pink skin, yellow flesh and outstanding flavour.

'Princess' (AGM) – Latest new potato, already a prize winner. Oval tubers, cream-coloured, floury flesh. Good flavour.

'Ratte' (AGM) – Introduced 1872. Highly popular in France. Apart from having a more normal appearance, it has all the good qualities of 'Pink Fir Apple' and is more productive.

'Roseval' (AGM) – An early maincrop potato. The deep red skins keep much of their colour when cooked and contrast well with yellow flesh. Unusual ruby foliage. Particularly popular in France. Great flavour.

'Sarpo Mira' – A wonder potato with near-complete blight resistance.

SECOND EARLIES

'Catriona' – A favourite for exhibition with blue-purple eye splashes. Needs good growing conditions.

'Dark Red Norland' (AGM) – Heavy cropping. Red skins with white flesh. Good for boiling and roasting.

'Kestrel' – Good disease resistance, and the slugs don't like it. Dominates show classes.

'Kondor' (AGM) – Introduced 1984. Red-skinned and waxy. Good taste and high yields. Good for baking. Fairly disease resistant.

'Lady Christi' (AGM) – Nearly a first early as it bulks up quickly. Medium sized, waxy, with creamy flesh. Foliage blight-susceptible but tubers are blight- and eelworm-resistant. GW recommends it as "a favourite early variety."

'Maxine' (AGM) – Introduced 1993. Five-star, highly popular, red waxy potato. Excellent for showing. Productive with high eelworm resistance.

'Nadine' (AGM) – Introduced 1987 and is grown worldwide. Heavy yielding. Creamy, waxy flesh that doesn't discolour when cooked.

MAINCROP

'Avalanche' (AGM) – Introduced 1989. Good for mashed potato, chips, baking. Good yield and flavour.

'King Edward' – perhaps the most famous potato of all.

'Maxine' (AGM) – Introduced 1993. Five-star, highly popular, red waxy potato. Excellent for showing. Productive with high eelworm resistance.

'Navan' (AGM) – Heavy cropper. White skin with creamy white flesh. Good flavour, excellent for frying. Eelworm resistant.

★*'Picasso' (AGM)* – Introduced 1992. Very high yielding. Creamy skin and bright red eyes. Waxy flesh. Excellent when boiled. Similar to 'King Edward'. Good eelworm resistance, and some resistance to scab. Stores well.

'Remarka' – Makes tasty jacket potatoes. Good all-round disease resistance.

'Sarpo Axona' – A breakthrough with near-perfect resistance to blight and good resistance to slugs and wireworm. Medium to large red-skinned tubers. High yielding, suitable for cooking in all ways, though particularly good for roasting and jackets. Good taste and stores well.

MINI POTATOES

★*'Mimi' (AGM)* – Plentiful, tiny new potatoes, of uniform size.

Seed potato to harvest: first earlies 100 days; second earlies 110–120 days; maincrop 140 days.

LEGUMES

(PODDED VEGETABLES)

Historically, the pea and bean tribe were valued above all as a winter food as they can be dried and stored, or ground into flour. They have nitrogen-fixing nodules on their roots so, if you leave the roots in the ground to rot down, they will enrich the soil. Runner beans, French beans and broad beans are amongst the most popular allotment crops, being easy to grow and prolific. The more you pick, the more will come.

French and runner beans are tender and will twine around a support by themselves. Peas and broad beans are hardy, cool weather crops. Peas cling on by their tendrils and will happily scramble up a row of pea sticks, hazel twigs, or netting supported by stakes. The supports should go in before planting.

Soil and situation

All need sun to do well. Apart from clay-loving broad beans, legumes prefer a light but rich soil on the alkaline side (pH7–7.5). Organic matter should be well mixed in before planting or you can grow them on a 'bean trench'.

Cultivation

To give them the best start and avoid the mouse and pigeon problem, sow the seeds under cover in a piece of guttering. This is a neat trick as you just slide them off the guttering when transplanting without disturbing the roots. They can also be station sown outside with cloche protection. Sow three or four per station.

Once transplanted, keep the plants moist and protect from weeds with mulch. Avoid watering too much as this will encourage growth in the leaves rather than the flowers and seeds, but water generously when they start to flower to maximize the crops.

Problems

Rotation is important to avoid a build-up foot and root rots. Birds (particularly pigeons), rodents, slugs and snails can be a problem. Broad beans are prone to black bean aphid. Pea and bean weevil, root aphid and red spider mite can occur. French beans can get anthracnose and halo blight.

BROAD BEANS

Vicia faba

The broad bean was recorded by Neolithic man, the Greeks, the Egyptians and Romans and it was mentioned in the Bible. It was a staple food for both rich and poor prior to the arrival of the potato, and the only bean widely grown in the UK until the 20th century. Its near cousin, *Vicia faba* var. *equina* was grown as horse fodder – hence the expression "full of beans".

Broad beans are the hardiest of the legumes and very easy to grow. They are an ideal crop for the allotment where there is usually a good open situation and no shortage of space. The flowers are sweetly scented.

Broad bean types

Types vary from dwarf to tall, kidney-shaped to round, and from green and nearly white to red. Opinions vary as to which are the tastier. The Longpod types with eight seeds are hardier and more prolific – the main choice for early crops. The shorter Windsors (first cultivated by Dutch gardeners at Windsor) have four seeds. They are considered the finer beans and are generally grown later in the season. Breeders have combined the two for new varieties with the merits of both. The dwarf types are usually sown for late summer harvest.

Cooking

The entire pods can be eaten raw or cooked like mangetout if they are picked when tiny. The French cook the mature beans with savory. The Chinese serve them in their skins to be shucked at the table. In Morocco they make purées of broad beans, combined with other pulses. The Egyptians make *ful medames* from dried broad beans.

Soil and situation

Broad beans are not fussy, though they prefer a heavy soil. They cannot cope with waterlogging or drying out in summer. It's important to dig the soil over well so their hefty tap root can grow down through it with ease.

Cultivation

The hardiest varieties can be grown from mid-January into February under cloches once the soil temperature has reached a minimum 5°C (41°F). Soak the seeds over night. For a continuous supply, sow once a month from then onwards, switching to main-crop types from March to May for beans throughout summer. In mild areas you can sow in November to overwinter, though this is not without risk.

Sow single seeds 5 cm (2 in) deep and 20 cm (8 in) apart. They are usually sown in a staggered double row with a good gap between rows 75 cm (30 in) so they don't cast shade on each other.

The easiest way to support broad beans is to run stakes along the row about 1 m (3 ft) apart and tie the tops round with string or wire.

Watch for mice as they may well be after the seeds before they even germinate! Keep the ground around them well hoed. Remove all suckers as they appear so that you are left with a single stem for each plant. When the tops have four clusters of flowers, cut them off. This will encourage the pods to form and help to protect against black bean aphids that are drawn like a magnet to the fresh young growth. Keep watered in dry spells.

Harvesting

Harvest as soon as the pods are ripe and before they coarsen.

Problems

Mice, black bean aphid, chocolate spot. Sometimes rust can appear but does little harm.

Recommended varieties

★'*Aquadulce Claudia*' (*AGM*) – A very hardy Longpod suitable for autumn planting. A top favourite since 1844. Compact plant. The Aquadulce series, also including 'Super Aquadulce', are the hardiest type of broad bean.

'*Express*' (*AGM*) – A strong hardy plant with mild-tasting beans good for freezing – the quickest for spring planting. Well-filled pods.

'*Imperial Green Longpod*' (*AGM*) – Another good one for freezing. It is noted for its flavour.

'*Masterpiece Green Longpod*' – Slim but well-filled pods also noted for flavour.

★'*Red Epicure*' – Red seeds and crimson flowers.

★'*The Sutton*' (*AGM*) – A dwarf variety, up to 45 cm (18 in), so good for windy sites. Excellent flavour.

★'*Witkiem Manita*' (*AGM*) – One of the earliest for spring sowing, with high yields, good flavour and uniformity. Not too tall. GW recommends it as "best from spring sowing."

'*Witkiem Major*' (*AGM*) – Has similar good qualities to 'Witkiem Manita'.

Seed to harvest: spring-sown 12–16 weeks; autumn-sown 28–35 weeks.

FRENCH BEANS

(FLAGEOLET AND HARICOT, KIDNEY, BUSH, POLE, SNAP,)

Phaseolus vulgaris

French beans, grown in Peru from 6,000 BC, were introduced to Europe during the Spanish conquest in the 16th century. They subsequently became known as 'French' beans as new strains were mostly imported from France in the early days.

An easy and prolific crop, they are a truly delicious when fresh and young. If for some reason you forget to pick them, all is not lost. They will mature into flageolet beans and then into haricot beans for drying and storing. As they do a triple act they are aptly known in China as *sandomame* or the 'three times bean'.

These fast-growing annuals need warm conditions.

French bean types

The two main types are climbing and bush. The climbing varieties will cling and can be grown up supports in the same way as runner beans. The dwarf types make low bushes which have the advantage of fitting neatly under cloches in cooler areas.

To add to the fun, consider unusual colours – red, purple, yellow and flecked as well as the usual green. There are types with pretty purple, lilac or white flowers. The beans come flat or pencil-shaped. Modern cultivars are usually stringless.

Cooking

When tiny, French beans can be lightly steamed and served hot with butter and herbs, or cold with dressings. As they mature they can be stewed in the Greek way with onions and tomatoes.

Flageolet beans are shelled, cooked and eaten like peas, but they take longer to cook.

Dried haricot beans are soaked overnight and simmered slowly until tender for stews, soups and bean salads.

Soil and situation

Choose a sunny and sheltered site. For good results don't attempt to grow them before the soil has reached 13°C (55°F). They like a light and fertile soil, near neutral (pH7).

Sowing and planting

Sow in late spring, having warmed the soil if necessary by covering with polythene for a couple of weeks. French beans are usually sown in staggered double rows for extra warmth. There should be at least 60 cm (24 in) in between rows for ease of picking.

Make parallel drills 5 cm (2 in) deep and station-sow two seeds, scar downwards,

POTATOES see pages 10–14

'Accent'

'Foremost'

'Duke of York'

'Cara'

'Pink Fir Apple'

'Charlotte'

'Catriona'

'Kondor'

'King Edward'

'Ratte'

'Lady Christi'

'Mimi'

'Picasso'

'Romano'

LEGUMES see pages 14–30

Broad bean 'Aquadulce Claudia'

Broad bean 'The Sutton'

Broad bean 'Witkiem Manita'

Broad bean 'Red Epicure'

French bean 'Eva'

French bean 'Brown Dutch'

French bean 'Barlotta Lingua di Fuoco'

French bean 'Hunter'

French bean 'Mont d'Or'

French bean 'Purple Queen'

French bean 'Atlanta'

French bean 'Mini'

Runner bean 'Painted Lady'

Runner bean 'Scarlet Emporer'

Runner bean 'Desirée'

Runner bean 'Hestia'

Runner bean 'White Lady'

Runner bean 'Best of All'

Runner bean 'Pickwick'

Runner bean 'Enorma'

French bean 'Safari'

Sugar snap peas 'Sugar Anne'

Sugar snap peas 'Sugar Lord'

Mangetout 'Carouby de Maussane'

Peas 'Kelvedon Wonder'

Peas 'Ezettas'

Peas 'Hurst Greenshaft'

Asparagus pea

every 15 cm (6 in) for early crops and 23 cm (9 in) for main crops. Sow a few extras as the germination rate of French beans is only 75 per cent. For a continuous supply, sow a few seeds every two or three weeks until July.

Cultivation

Climbing French beans are self-twining and will scramble up twiggy pea sticks, netting, a pole or even up sweetcorn. The bush types are earthed up to the first set of leaves to give them extra support. Protect against slugs, birds and mice. Keep the plants moist and well mulched. Water copiously when in flower.

Harvesting

French beans will continue to produce if you keep gathering the beans when young. They are ready when they snap off. Flageolet beans have to be caught at the intermediate stage, when the seeds are quite small and still bright green.

For haricot beans, pick each pod as it ripens. If the weather turns cold, pull out the entire plant at the end of the season and hang it out to dry in an airy place. Shell haricots when the skins are dry and crackly. Dry the beans further for a couple of days in a sunny room or in the airing cupboard and store in airtight jars.

Problems

Most common are slugs and aphids. Anthracnose and halo blight can occur as well as foot rots.

Recommended varieties

CLIMBERS

'Algarve' (AGM) – Stringless slicing beans up to 25 cm (10 in) long. Good, consistent cropper with excellent flavour.

**'Eva' (AGM)* and *'Diamont' (AGM)* – Early varieties with long, round pods and black seeds. Resistant to bean mosaic virus.

**'Hunter' (AGM)* – Strong-growing, heavy cropper with straight stringless pods, some 23 cm (9 in) in length. White seeds. Popular for exhibition.

'Kingston Gold' (AGM) – Yellow pods.

FOR DRYING

**'Barlotta Lingua di Fuocco'* – The Italian 'Fire tongue' bean has bright green, flat pods with red markings (which disappear when cooked). A delicious bean.

**'Brown Dutch'* – Brown beans.

'Horsehead' – Dark red beans.

MINI VARIETIES

'Aroza' – Resistant to mosaic virus.

'Cantare (AGM) – Early, high-yielding, good flavour and resistant to bean mosaic virus.

'Cropper Tepee' (AGM) – Pencil podded cultivar. High yield. 'Purple Tepee' is the purple version.

**'Mont d'Or'* – Yellow pods.

'The Prince' (AGM) – Produces masses of delicious slender flat pods. Good for exhibition.

**'Purple Queen'* – Purple pods.

'Sprite' (AGM) – Stringless Continental variety with dark green round pods. Heavy yields.

MINI VARIETIES FOR DRYING

'Barlotto Lingua di Fuocco Nana' – Bright green, flat pods with red markings.

DISEASE-RESISTANT VARIETIES

'Copper Tepee' – Resistant to anthracnose.

'Daisy' – Resistant to common bean mosaic virus.

'Forum' – Resistant to halo blight, anthracnose and common bean mosaic virus.

Seed to harvest: 7–12 weeks.

RUNNER BEANS

Phaseolus coccineus

Runner beans were introduced from Mexico by John Tradescant in the 17th century. A very elegant plant, they were grown as ornamentals until the

18th century when it was discovered that they were good to eat. They are frost-tender, vigorous climbers, growing to 2.5 m (8 ft) or more.

Runner bean types

There are climbing and dwarf varieties. The dwarf varieties are good for early crops under glass but are not so prolific. The flowers are usually red, but there are also white and two-tone ones. Modern breeding has concentrated on producing beans that are stringless and less fibrous.

Cooking

As for French beans. Classically they are cut into fine ribbons and lightly cooked as a side vegetable.

Soil and situation

Runner beans need a good root run, some 38 cm (15 in) of top soil. It should be fertile, deep, moisture-retentive but free-draining soil, pH6–7. Prepare it in the autumn by digging it over and enriching it with plenty of well-rotted compost. They need a sunny site though they don't like to be baked at midday. It should be sheltered to allow the optimum conditions for bees to pollinate the plants.

Sowing and planting

Even a whiff of frost will kill runner beans, so wait until May or June before sowing. Alternatively, seed off indoors in root trainers or biodegradable pots at a minimum temperature of 12°C (52°F).

Set up strong supports before sowing. The traditional method is a row of poles tied at the crossover point with wire or string, and anchored at each end by a short stake.

Make a V-shaped trench 5 cm (2 in) deep. Sow the seeds 2.5 cm (1 in) deeper and fill the trench to the top when they germinate to protect them from cold. If sowing two staggered rows, space the seeds about 30 cm (12 in) apart in rows 38 cm (15 in) apart. Sow a few extras as the germination rate is only 80 per cent.

Cultivation

Mulch well and keep moist. Tie in the young plants until they are able to twine. Pinch out the tips when they reach the top of their supports to prevent them getting top heavy. As the flowers appear, give the plants a thorough soaking.

Harvesting

Pick the beans as soon as they are ready. Check through the foliage carefully to make sure that there are no old pods lurking there as they will stop production. Runners usually produce a bumper crop over many weeks.

Problems

Usually trouble free. They can suffer from poor setting when, despite the fact that the plant appears to be healthy, the beans are distorted or have dry patches. This can be caused by frost, bad weather or dry soil.

Recommended varieties

**'Desirée'* and *'Lady Di' (AGM)* – Both almost stringless, and prolific croppers. 'Lady Di' has long, slim, blemish-free pods.

**'Enorma' (AGM)* – Big cropper. Long, slender beans, good for showing.

**'Painted Lady'* – Introduced in 1855, it is the oldest variety of runner bean. Exceptionally pretty with scarlet and white flowers, it is also known as 'York and Lancaster'.

'Red Flame' – Another new stringless type with red flowers.

'Red Rum' (AGM) – Early, with little foliage but masses of beans. Tolerant of halo blight.

**'Scarlet Emperor'* – Introduced in 1906, it remains a great favourite. It is said by many to be unbeatable for flavour.

'White Emergo' (AGM) – Prolific late cropper with tradi-

tional taste and texture. It is known for its vigour, which helps it to cope with bad weather. White flowers and seeds.

★*'White Lady' (AGM)* – A new stringless variety with white flowers. It is very prolific. Said to be less prone to birds and good at coping with high temperatures. GW gives it top marks for flavour.

MINI VARIETIES

★*'Hestia'* – A new variety which stands clear of the ground, is easy to pick and needs no support. Disease-resistant. Heavy cropper with red and white flowers. Grows to 45 cm (18 in).

★*'Pickwick'* – A modern, early dwarf variety growing no higher than 60 cm (24 in). The beans are stringless if picked young.

Seed to harvest: 8–12 weeks.

LABLAB BEANS

(HYACINTH BEAN OR DOLICHOS BEAN)

Lablab purpureus

This is a gorgeous plant with rose-coloured flowers superb for cutting, black beans, and big heart-shaped leaves along with many variations of a theme. In China, where it has been grown for centuries, it grows to 6 m (20 ft) and fruits prolifically. A perennial, it will reach 1.8 m (6 ft) in the UK if provided with the warmth it needs and if treated as an annual. For a guaranteed crop of beans, however, it should be grown in a polytunnel.

Cooking

As for French beans.

Soil and situation

The lablab bean can cope with poor soil, acid or alkaline, but requires good drainage and the warmest and most sheltered place on offer.

Sowing and planting

In late spring, soak seed overnight and germinate on damp kitchen towel. Start off in a propagator at 15°C (59°F). Harden off and transplant when the soil temperature is a minimum 15°C (59°F).

Cultivation

Normal care. Once established, lablab beans are very tolerant.

Harvesting

Pick young beans as soon as they are ready. Don't dry the seed to store as it will develop toxic cyanogenic glucosides which can only be removed by many hours of boiling and changes of water.

Problems

As for all legumes.

Recommended varieties

None in the UK.

Seed to harvest: in the tropics, seed sown in summer will flower in both autumn and spring. In the UK, seed sown in spring will produce flowers in late summer and beans after if conditions are favourable.

YARD-LONG BEANS

(CHINESE LONG BEAN, SNAKE BEAN, ASPARAGUS BEAN, GARTER BEAN, BLACK-EYE BEAN, CROWDER)

Vigna unguiculata subsp. *sesquipedelis*

The yard-long bean is an ancient Asian plant growing up to 3.6 m (12 ft) high. True to its name, this bean grows up to a yard long in the optimum conditions of its native climate. It is usually harvested when the beans are about 30 cm (12 in). It has pretty flowers that open in the morning and close at midday. Today it is grown widely in the warmer parts of the USA as well as in India and West Africa.

It's an unlikely proposition for outside on the allotment in our cool, temperate climate. The traditional types take three or four months to mature. If you can get hold of them, a better bet for the UK are the faster-maturing new cultivars. In warmer parts of the country with coverings, there is no reason why it should not be grown at least as a novelty. No doubt the breeders will come up with some hardier types in due course.

Yard-long bean types

There are climbers and dwarf types. Climbers won't reach anything like their full potential in the UK. Dwarf varieties are easier to keep warm.

Cooking

The beans are less sweet than runners and are usually chopped into short lengths for stir-fries. They go mushy if cooked for too long. The purple varieties keep their colour when cooked.

Soil and situation

Choose a well-drained, light, fertile soil and a sheltered hot spot.

Sowing and planting

Sow seeds in a heated propagator set to 18°C (64°F), 5 cm (2 in) deep. Germination will take place in 6–10 days. Pot on, avoiding disturbance to the roots and taking great care to keep the plants protected from any whiff of cold. Plant out under fleece or in a polytunnel when the minimum temperature is set fair at 20°C (68°F). The climbing varieties need staking.

Cultivation

Keep well watered. If you are growing under fleece or cover, open it up on warm days when the plant is in flower for pollination. Modern fast-growing cultivars may flower five weeks after sowing and fruit about two weeks later.

Harvesting

Pick the beans while young and tender and about 30 cm (12 in) long.

Problems

Red spider mite due to heated conditions.

Recommended varieties

Available in the USA are ***'Red Noodle'*** and ***'White Seeded'***.

Seed to harvest: from 7 weeks.

PEAS

(GARDEN PEAS, PETITS POIS, MANGETOUT AND SUGARSNAPS)

Pisum sativum

Peas are said to be the oldest cultivated vegetables in the world. Seeds were found in the Bronze Age dwellings of the Swiss Lakes and in the ruins of Troy. Breeding in France and Holland transformed the bitter wild pea into the popular vegetable we know today. Four varieties of mangetout, including 'Pease without skins in the cods', were mentioned in Gerard's 16th century *Herball*. However, peas are quite a challenge to grow well. To get good servings of peas you need quite a few plants. It is reckoned that a 5 m (17 ft) row will yield four pickings of 1 kg (2 lb) of peas.

Pea types

There are garden peas, petits pois, peas for drying, mangetout with flat pods, and the sugarsnap types with swollen pods. The leafless peas have been developed for mechanical harvesting and need no support. Round-seeded varieties are used for the cooler conditions of spring and autumn and early summer. The sweeter but less hardy wrinkled types are designed for maincrop sowings.

Cooking

Fresh peas are a summer delicacy cooked briefly with mint. In France they are gently stewed with a lettuce heart, spring onions, parsley and butter for *petits pois à la française*. In Italy, the young peas are enjoyed in *risotto primavera*.

As the name implies, mangetout are eaten pod and all, as are sugarsnap peas. They are good stir-fried with ginger and spring onion, particularly the young shoots of the leafless peas.

Dried peas are traditionally combined with boiled ham for soup or pease pudding. Peas contain protein, carbohydrates, iron and vitamin C.

Soil and situation

They like a fertile, light, moisture-retentive soil. Prepare the bed in the autumn, adding plenty of well-rotted compost or manure. Peas need a good root run and access to constant moisture. They can take a little shade.

Sowing and planting

For peas in May or June, earlies can be sown under cloches in soil at a minimum 7°C (46°F) in late February or March. As peas are hardy but the flowers are not, it is safer to germinate them on damp kitchen towel and then pot on into modules. Harden them off and then transplant into a piece of guttering and grow on until the weather is warmer in a greenhouse or polytunnel.

Carry on sowing seeds every couple of weeks if you want peas throughout the season. Change to wrinkled types and round second-earlies in mid-March to mid-April. The second earlies usually produce a much better crop than the first.

For peas in August you move onto maincrop wrinkled types. These are taller, slower but are the best quality if grown well.

For autumn peas in mild areas, make a last sowing in July with fast-growing early round types which have mildew resistance. Some new cultivars can be sown at any time.

If planting outside, prepare a trench and sow them individually about 5 cm (2 in) deep and quite close together, at about 7.5 cm (3 in), to allow for losses. The distance between rows is calculated to be roughly the same as the ultimate height of the particular variety of pea.

Cultivation

Mulch to keep in moisture, leaving a little gully between rows to collect rain water. Protect them against mice and birds from the moment they go in with wire mesh.

When the peas have grown to about 7.5 cm (3 in) and the first tendrils appear, put a little brushwood around them for the pea tendrils to get a hold and start to climb. Remove the wire mesh at this stage. Keep moist and give them plenty of water, preferably rain water, when in flower.

Harvesting

Test a few podded peas to see if they are ready. Pick regularly, supporting the stem as you do so. Mangetout and sugarsnap peas are harvested when the peas are just visible in the pod. If you neglect to catch them at this stage you can use them in the same way as ordinary peas. They are best eaten as fresh as possible.

Problems

Pea weevil, aphids, pea moth, birds, mice and slugs.

Recommended varieties

PEAS

'Cavalier' (AGM) – Long pods, late, sweet taste. Doesn't mature all at once. Mildew resistant.

'Early Onward' (AGM) – Early, sweet tasting and prolific.

★'Kelvedon Wonder' (AGM) – An old English favourite, still going strong. Dwarf type with large, well-filled, mid-green pods. Good for successive sowing.

'Saturn' (AGM) – Maincrop. Very fruitful with a long picking period.

'Show Perfection' – High-yielding maincrop exhibition pea.

SUGARSNAPS

★'Sugar Anne' (AGM) – Very early to mature. Huge crop of pale green pods with a sweet flavour.

★'Sugar Lord' (AGM) – Tall and vigorous. High yielder.

MANGETOUT
★'Carouby de Maussane' – Ornamental flowers and large, flat pods. GW recommends it.
'Delikata' (AGM) – Tall, similar to 'Oregan Sugar Pod' though a little earlier. If not picked regularly, the pods will get strings. Mildew and fusarium resistant.
'Oregon Sugar Pod' (AGM) – Extended picking period of medium flat pods which should be picked when young and stringless. Resistant to powdery mildew, common wilt and virus. GW describes it as "a good tall mangetout."
'Snow Wind' (AGM) – Semi-leafless with dark green, sweet-tasting pods. Crops well over a long season.

MINI SUGARSNAPS
'Cascadia' (AGM) – Dwarf habit. Heavy cropper. Pods stay tender and sweet for a long time.
'Delikett' Dwarf' (AGM) – Stringless when young. Fleshier and sweeter as matures. GW describes it as "the best snap pea".

MINI MANGETOUT
'Edula' (AGM) – Fairly dwarf and a good cropper.

Seed to harvest: earlies 11–12 weeks; second earlies 12–13 weeks; maincrop 12–13 weeks.

ASPARAGUS PEA

Lotus tetragonolobus

The asparagus pea is something of a rarity – not really a pea at all but a small creeping vetch, no higher than 15 cm (6 in) and 60 cm (24 in) wide. It's not a great cropper but is so pretty that perhaps it is worth growing for its enchanting red flowers. The winged pods are a delicacy as long as they are picked when young. A Mediterranean plant, it can take more heat and dryness than the ordinary pea. It will enjoy a sunny spot in light but fertile soil. Sow seeds under cover in April or May, barely covering the seed. Once hardened off they will be ready to plant out six weeks later for eating 8–10 weeks after that.

Recommended varieties

Available in British catalogues as ★'Asparagus Pea'.

Seed to harvest: 14–15 weeks.

BRASSICAS

Primitive forms of cabbage grow wild on the chalky coastlines of southern England, through most of Europe and as far as North Africa. They are the forebears of the largest of all the vegetable families, the *Brassicaceae*. A tribe of nutritious, cool-climate crops they all have much the same needs. Being leafy, they need plenty of water in dry spells. A good soak before harvest will make a big difference to the crop. Some are sown outside where they are to grow, as they are likely to bolt when transplanted. Others – particularly the slow-growing ones – are generally started in the seedbed or in modules for a more controlled, pest-free start and to save space.

Many brassicas grow away again after the first cut to send out a second harvest of tasty green tops. Summer sowings are harvested when ready, but the winter ones usually stay in the ground until they are needed. The soil needs to be fertile, well-drained but moisture-retentive. It should be prepared well in advance so it has time to settle down. Land that has been manured for the previous crop is ideal. Brassicas like firm ground.

Clubroot is their worst enemy, but they can also suffer from downy and powdery mildew as well as bacterial leaf spot. Protect them from cabbage root fly with collars of rubber-backed carpet underlay, the commercial equivalent, or by growing under fine netting or mesh. Watch out for cabbage moths, cabbage white butterflies, mealy cabbage aphids, cabbage whitefly, cutworms, birds and slugs and snails. Flea beetle might appear in dry weather.

BRUSSELS SPROUTS

Brassica oleracea Gemmifera Group

It would seem that the Brussels sprout arrived from Belgium in the late 18th century. Sprouts are extremely hardy (the toughest can cheerfully withstand –10°C (14°F) and they can be picked fresh over a period of two to three months. Slow growing and columnar in shape, they are good candidates for undercropping with salad leaves or faster-growing greens.

Brussels sprout types

Sprouts are classified according to season. There are also red or purple varieties. The F1 hybrids have been bred to avoid the traditional problems, especially the habit of producing "leafy blowers" instead of tight sprouts. They produce good quantities of uniformly tightly packed buttons. They are also stockier plants, so are less likely to get blown over, and are sweeter tasting.

Cooking

If you put the freshly picked sprouts in salted water, any hidden insect life will emerge. Sprouts are generally cooked until *al dente*, and served with butter. They are a good source of vitamins A and C, as well as iron. For the full benefit, eat them as fresh as possible.

Soil and situation

Sprouts like classic brassica soil (see above). Make sure that it is firm to secure them from wind rock. They can take a little shade and even do well on a north-facing slope.

Sowing and planting

Earlies For sprouts at Christmas, sow in early spring outside under cloches. Sow thinly, 1 cm (1/2 in) deep. Cover with a cloche mounted on tiles or stones to let in air. Keep the seeds in the dark until they germinate. After that take the cloche off on sunny days, replacing it at night or when it rains heavily. Be ready to replace it with covers if the weather turns cold. When big enough to handle, thin out to 15 cm (6 in) apart. Transplant in May, when plants are about 23 cm (9 in) high, taking as much soil with the plants as possible. Arrange them 60 cm (24 in) apart both ways.

Even earlier sprouts for autumn can be germinated with bottom heat of 18°C (64°F) in a propagator in late winter. Remove them from the propagator fairly smartly after about a week and keep them at a cool 10°C (50°F) to prevent them getting leggy. Transplant into modules when big enough to handle to encourage a strong taproot to form. Plant out in spring.

Late sprouts Sprouts ready in the New Year and onwards are started off in April and grown in the same way as earlies, without the weather worries.

Cultivation

Once in their final positions, sprouts need little attention. Keep them weed free, watered in dry spells and staked on the windward side as they get bigger. If the sprouts are growing unevenly and you want to crop them all at once for freezing, earlies (but not lates) can be stopped. Pick off the sprouts at the tip of the plant in June or July. In autumn, stabilize the plants further by earthing them up to the level of the first set of leaves.

Harvesting

Remove any yellowing or diseased leaves regularly. Harvest from the bottom upwards. Dig out entirely at the end of the season but wait until they have made one final departing gesture by throwing out a tasty top shoot.

Problems

Watch for mealy aphids as they can get right inside the sprouts.

Recommended varieties

EARLY

'Diablo' (AGM) – Early to mid-season. Good-quality, clean, round sprouts of good flavour. Recommended by GW.

'Icarus' (AGM) – Large, smooth, solid sprouts. Plants stand up well. Susceptible to white blister and ringspot.

MID-SEASON

'Cavalier' F1 – Dark green, well-spaced sprouts. Plants produce high yields of good quality.

'Clodius' (AGM) – Good-quality round, smooth, solid, sweet sprouts. Plants stand and yield well.

****'Igor' (AGM)*** – Midseason to late. Small smooth sprouts which tend to become loose if not picked in time.

'Lunet' F1 – Tall plants, producing a very high yield of large, oval, fairly sweet sprouts, which are easy to pick and have good resistance to ringspot. Plants can lean over.

'Patent' (AGM) – Large, round, dark green solid sprouts.

'Roger' (AGM) – Large smooth, good-quality pale green sprouts. Plants stand well.

LATE

'Bosworth' (AGM) – Oval, mid to dark green, solid, closely spaced sprouts. Easy to pick; plants stand well. Quality still good in February. GW describes them as sweet tasting.

'Cascade' (AGM) – Smooth, clean, well-spaced, dark green button sprouts. Uniform plants that stand and yield well.

'Millenium' – Recommended by GW for sprouts into the New Year.

'Noisette' – A nutty, French variety that can crop from October to February.

'Wellington' (AGM) – Very good-quality, smooth, dark green, round, solid sprouts.

MINI VARIETIES

****'Peter Gynt'*** – Tried and tested F1 dwarf variety. An early starter with a long cropping season.

'Rubine' – Fairly dwarf with tasty deep red sprouts.

Seed to harvest: 20 weeks.

CABBAGE

Brassica oleracea Capitata Group

The bitter wild cabbage, that grows on the seacoasts of northern Europe, was eaten by the Celts before the Roman invasion. Most of the cabbages that we know today were cultivated in Germany from the 12th century onwards. The range nowadays is so wide as to be bewildering. There are eight main types and myriad introductions and crosses. Some seed merchants offer collections – an economic way to see what suits you. Alternatively, share and exchange seed with your allotment neighbours to ring the changes. It is always a good idea to stagger sowing over a couple of weeks to optimize your chances and prevent a glut.

Cabbages are not hard to grow as long as you provide them with good soil, continuous moisture and take evasive action against numerous pests and diseases.

Cabbage types

The main types are usually classified by their harvesting times – spring, summer and autumn. The winter ones are divided into winter whites, January King cultivars, savoys, savoy hybrids and red cabbage. The winter whites are mild tasting and store throughout winter. The savoys and the January King cultivars are bred to stand out all winter. The savoy hybrids, a cross between the savoy and the winter white, have the hardiness of one parent and the milder taste of the second. The red cabbage is harvested in late autumn and can be stored for a couple of months.

Cooking

The winter whites are the classic cabbage chopped raw for coleslaw, though the crinkly savoys are also excellent. Red cabbage is usually cooked with apple and the others are used as greens – steamed, braised, boiled, pickled, stir-fried or wrapped around fillings in the Greek way for *dolmades.*

Soil and situation

Cabbages like fertile, well drained soil and an open situation. To help their roots to get a good hold, use a dibber to plant them out and firm the soil round them well. If your soil is light, make a shallow drill and earth the young cabbages up as they grow.

Cultivation

Spring cabbage Spring cabbage has a dual role – to provide spring greens in March or April and hearted cabbage in April, May and June. Make two sowings a couple of weeks apart at the end of July and in August to hedge your bets. You are aiming to have young plants that are sturdy enough to withstand the frosts but not so advanced as to bolt in the first sunshine of spring.

Either sow spring cabbage where you want it to grow or in modules, transplanting the seedlings in late September or October. If you plant them out 60 cm (24 in) apart, the second set can go in between. You can use whichever set is less successful as spring greens while the other hearts up in the extra space. Cloches can be used to speed up part of the crop, or stagger it. Don't fertilize spring cabbage through the cold of winter as it will encourage a flush of soft growth susceptible to frosts.

Summer cabbage This can carry on right through summer if you start with early varieties and move onto main-crop types. Earliest crops are sown in March outside under fleece for June eating. For even earlier crops, start them off in a propagator at 16°C (61°F) in February. Harden them off well before transplanting on a cool day, at around 20°C (68°F) to prevent bolting. For a succession of summer cabbage, sow more seeds every couple of weeks through April. A final sowing at the end of April will give you cabbages for September. Space 45 cm (18 in) all round.

Autumn cabbage Change varieties and sow outside from May onwards for autumn to early winter eating.

Winter cabbages These are sown in late spring to mature through the winter into the following New Year. Space about 45 cm (18 in) apart.

Harvesting

Spring and summer cabbages are always eaten freshly harvested. Spring cabbages are picked as greens while the others are harvested when they have a solid heart. If the stumps are left in the ground, they may produce some tasty top shoots, especially if you make a cross cut on them. Winter whites, as well as some red cabbages, store for up to five months.

Dig them up carefully to avoid damage before the frosts and hang (either with all the roots or a good-sized stump) in nets or pack loosely in straw.

Check from time to time and remove any damaged outer leaves. Savoys and the January King cultivars are cropped as needed into the New Year. At the end of the season, completely clear the ground.

Problems

Prone to all brassica pests and diseases.

Recommended varieties

SPRING CABBAGE

**'Duncan' F1 (AGM)* – Both for spring greens and hearted cabbage as they heart up slowly, producing small, solid heads. A good early yield.

'First Early Market 218' (AGM) – Good colour and conformity. Hearts up slowly to produce large, quite loose heads.

'First Early Market 218 – Mastercut' (AGM) – Attractive dark green outer leaves; well filled hearts.

**'Offenham 1 Myatt's Offenham Compacta' (AGM)* – Big-hearted cabbage.

'Offenham 3 – Mastergreen' (AGM) – Uniform, bright green spring greens.

SUMMER CABBAGE

'Augustor' F1 (AGM) – Bright green, compact, uniform with solid round heads. Mid-season.

'Charmont' F1 (AGM) – Round type, bright green.

'Derby Day' F1 (AGM) – Early, bright mid-green and round.

'First of June' F1 (AGM) – Early, round and mid-green.

**'Greyhound' F1 (AGM)* – Early, pointed type.

**'Hispi' F1 (AGM)* – Smooth and pointed with a solid heart.

**'Stonehead' F1 (AGM)* – Round type, mid-green. Can be cropped into autumn.

EARLY REDS

These varieties are not for storing.

'Lanfgedijker Red Early' (AGM) – Good solid heads.

'Rodeo' F1 (AGM) – Good colour.

'Rondy' F1 (AGM) – Ready early autumn.

'Ruby Ball' F1 (AGM) – Uniform round heads.

WINTER WHITE

'Marathon' F1 (AGM) – Stores well but needs to be harvested before the frosts.

SAVOY

'Alaska' F1 (AGM) – Mid-season to late. Dark green with good blistering. On the small side.

'Clarissa' F1 (AGM) – Early. Peppery taste.

'Famosa' F1 (AGM) – Very early. Bright green, big round heads with reddish veins.

'Midvoy' F1 (AGM) – Dark green blistered outer leaves, solid round hearts.

'Primavoy' F1 (AGM) – is earlier and *'Protovoy' F1 (AGM)* is earlier still.

'Tarvoy' F1 (AGM) – Mid-season to late and **'Wivoy' F1 (AGM)* is late.

SAVOY HYBRIDS

'Beretta' (AGM) – Dark green and blistered with mild-tasting hearts.

**'Celtic' (AGM)* – Large and leafy with solid heads.

'Embassy' (AGM) – Lightly blistered, bright green, sweet tasting.

'Renton' (AGM) – Early.

JANUARY KING

**'Flagship' F1 AGM* – Pronounced purple colour. Stands well.

**'Holly' F1* – Dark green with purplish tinge. Winters well.

'Marabel' F1 – Round heads, reddish, stands well.

MINI VARIETIES

'Golden Cross' F1 (AGM) – Small, round, uniform, bright green summer cabbage.

'Gonzales' F1 (AGM) – Compact round summer cabbage with pale leaves.

'Pixie' (AGM) – Very early hearted cabbage for spring.

'Robin' F1 (AGM) – Hybrid cabbage with dark green leaves, noticeably sweet and mild. Small and solid.

'Roulette' and *'Winchester' (AGM)* – Winter Savoys with small well-filled heads.

Seed to harvest: 20–35 weeks.

THE BROCCOLIS

Brassica oleracea Italica Group Early forms of broccoli were recorded by the naturalist Pliny the Elder. Catherine of Medici is thought to have introduced broccoli to France upon her marriage with Henri II. Miller's *Garden Dictionary* of 1724 describes it as "sprout colli-flower", but the name broccoli comes from *bracchium,* the Italian for a branch.

Broccoli types

Sprouting broccoli (*Brassica oleracea* Italica Group) is a slow-growing over-wintering plant and both the easiest and the most delectable of the group. As it takes up space for the best part of a year, it's not widely grown in gardens but is an excellent choice for the allotment. If small florets are picked frequently, it is reckoned that ten plants will give you about 3 kg (6 lb), so you don't need many. The purple forms are hardier and more prolific than the white ones.

Calabrese (from Calabria), the more usual type to be seen in the supermarket with a single head, or curd, is a summer crop. When the main head is harvested, the plants will send up succulent side shoots, each with a small curd and some leaves, for many weeks afterwards. Its pretty cousin, romanesco (presumably from Rome), is harvested in late summer or early autumn. Calabrese and romanesco are trickier to grow than sprouting broccoli as they have a tendency to bolt.

Perennial broccoli, or Nine Star, comes up every year, producing white cauliflower-like heads for harvesting in spring. It is usually replaced every three years.

Cooking

The broccolis can be lightly cooked as a vegetable accompaniment, used in soups or in the same way as cauliflower, with a cheese sauce. They are also good for stir-fries. The flowering stems are delicious steamed until *al dente* and served like asparagus with hollandaise sauce. Broccoli is highly nutritious, rich in vitamins C and A and contains the same amount of calcium as milk, weight for weight.

Soil and situation

The soil should be fertile and well-drained to avoid waterlogging in winter. The broccolis need shelter from the wind, especially sprouting broccoli as it can become top heavy.

Sowing and planting

Sprouting broccoli Sow in spring or early summer, either in modules or the seed bed, 2.5 cm (1 in) deep, 60 cm (24 in) apart, and 30 cm (12 in) between rows. The new cultivars that grow in a single season can be sown from late winter.

Calabrese Earliest crops can be sown in modules in late autumn under cloches and transplanted to a cold greenhouse or frame in winter. Otherwise sow in May in situ. It doesn't transplant well in the heat of summer.

Perennial broccoli Sow in mid-spring and transplant in early autumn 90 cm (3 ft) apart.

Cultivation

Keep weed free, watered in dry spells, earthed up, and firmly staked against winter gales. Avoid cosseting it with too much fertilizer or water as it needs to be brought up tough enough to cope with winter.

Harvesting

Pick sprigs when the buds are formed. Don't let them flower or production will halt. Cut the central stem first to encourage side shoots to grow. Keep picking these a few at a time over the next six weeks. The shoots dwindle in size towards the end of the season.

Problems

Net if pigeons take a fancy to them.

Recommended varieties

PURPLE SPROUTING

**'Bordeaux' (AGM)* – Early. Quite compact and stands well over winter.

**'Claret' F1 (AGM)* – Tall with a heavy yield of dark purple spears from March to April.

'Express Corona' – Good for resistance to downy mildew. Recommended by HDRA.

'Early Purple Sprouting Improved' – Ready to harvest in February or early March, a great bonus for the hungry gap. It's a well-tried, popular and traditional choice with a long cropping season.

**'Late Purple Sprouting' (AGM)* – For April picking. It's a good over-winterer and being late, it extends the season.

**'Red Arrow' F1 (AGM)* – Early mid-season, long cropping. Good winter hardiness, bushy and vigorous. HDRA describes it as "Quite short with masses of shoots." GW recommends it as far superior to older types.

Seed to harvest: 8–12 months.

CALABRESE

'Corvet' F1 (AGM) – A good summer crop.

'Decathlon' (AGM) – Does well on poor soils. GW recommends it for later but heavier heads.

'Flash' F1 (AGM) – Crops over a long period. Large heads that hold well.

'Hydra' F1 (AGM) – Mid to late vigorous variety; uniform. Average yield of medium-sized secondaries.

'Tenderstem' – A personal favourite. Produces an abundant crop of delicious spears.

'Tiara' F1 (AGM) – Sow in May for early medium-sized heads. GW recommends it for the earliest crop.

'Trixie' F1 (AGM) – Early mid-season. Blue-green leaves and attractive, medium-sized domed heads with small buds. High yield of secondaries produced 3–5 weeks after the primary heads. HDRA recommends it as resistant to club root.

HDRA recommends *'Emperor'* and *'Shogun'* for resistance to downy mildew.

Seed to harvest: 11–14 weeks.

CAULIFLOWER

Brassica oleracea Botrytis Group

Cauliflowers came from the eastern Mediterranean, probably from Arabia. Gerard's *Herball* of 1597 referred to them as "cole flowery" – perfectly logical as cauliflower is kale or cabbage bred for its flowerbuds. Cauliflowers didn't catch on in England until the 18th century when they became one of the few members of the cabbage family to be popular with the gentry. It must have been a headache for their gardeners as they are notoriously temperamental. New breeding has successfully overcome some of the problems.

Cauliflower types

Cauliflowers are classified by season, though some can be grown at any time. The early summer varieties, sown in autumn or winter and ready before the full heat of summer, are the easiest. Summer and autumn cauliflowers are grown in one season while winter types take the best part of a year. There are purple, green and orange varieties amongst the summer types. One to try is the pale green 'Floccoli', a cross between cauliflower and broccoli.

Cooking

Cauliflowers are natural partners with cheese and other creamy sauces. They are also eaten raw as crudités and in salads, either raw or blanched.

Soil and situation

Be particular with the cauliflower bed. Aim for top-quality, moisture-retentive soil. The autumn before planting, dig it over and lime (if necessary) to achieve a pH of 6.5–7.5. It should be well trodden down as cauliflowers hate loose soil. Sun, no wind and a constant level of moisture are essential for success.

Cultivation

Early summer types Sow seed in winter over a couple of weeks to hedge your bets and to avoid a glut. Use modules in the propagator set to 21°C (70°F). Harden off and transplant in early spring 45 cm (18 in) apart, with rows 60 cm (24 in) apart. Water the evening before you move them and firm them in thoroughly. Give them the protection of cloches or fleece. Alternatively, sow seed in autumn either in the seed bed or in situ.

Autumn cauliflowers These are sown in the same way in mid to late spring and transplanted in early summer. They must be kept very well watered in dry periods.

Winter and spring-headed cauliflowers These are sown in May in the seedbed and transplanted in July to a spot with frost protection. They need a cold spell. As the curds are not frost hardy, they are not worth trying unless your allotment is in a very mild, frost-free part of the country. Cover the curd with the leaves in summer to prevent it going yellow and in winter to protect it from harsh weather. Don't give them too much nitrogen as it encourages lush growth that will collapse in the cold.

Harvesting

Harvest as soon as they are ready. Don't cut off all the leaves as they will help to protect the curd. Cauliflowers can be stored for a couple of weeks if suspended by the roots and kept in a cool frost-free place.

Mini cauliflowers

The easiest way to grow cauliflowers is to keep them small and grow them quickly. Sow in April in situ for July eating. Space them 15 cm (6 in) apart both ways. Sow a few seeds every couple of weeks for a succession of crops.

Problems

The seeds often come up blind, without a central bud. Cauliflowers develop is in two stages – juvenile and mature. The juvenile stage is leaf growth. A period of cold weather triggers the mature stage when the white curds are formed. If these two stages are not synchronized, or the plants suffer any stress, buttoning may occur in the form of small curds poking out above the leaves.

Cauliflowers are sensitive to acidity and generally don't thrive in acid soil even when it has been limed. Acidity can cause 'whiptail' which shows as narrowing, mottling and yellowing of the leaves and also chlorosis, yellowing of the leaves. The curds sometimes turn brown from tipburn if the soil lacks sufficient calcium – a characteristic of acid soils.

If the soil is either too acid or too alkaline, the curds may go brown or rough patches can develop on the stems and stalks due to boron deficiency. If cold weather strikes when the curds are forming, the curds may develop 'riciness' – a strange reaction causing some of the flowers that form the curd to grow taller than the others.

Recommended varieties

SUMMER

'Nautilus' F1 (AGM) – RHS recommends these as "Vigorous plants with deep white well-protected curds of excellent quality."

'Perfection' F1 (AGM) – Very early. Large curds.

★'Plana' F1 (AGM) – Mid-season, uniform and vigorous.

'White Rock' F1 (AGM) – Late. White well-protected curds.

AUTUMN

'Alverdo' F1 (AGM) – Early to mid-season. Smooth green curds that look attractive when cooked with their contrasting white centre.

'Aviso' F1 (AGM) – An early cropper. Short cropping season but first-class, smooth white curds. Recommended by GW.

**'Esmeraldo' F1 (AGM)* – Smooth green curds.

**'Marmalade' F1 (AGM)* – Interesting variety with orange curds. Very early.

**'Minaret' F1 (AGM)* – Pyramid shape. Green curd.

**'Violet Queen' F1 (AGM)* – Very early with purple curds.

'Violetta Italia' F1 (AGM) – Very early with violet curds.

WINTER AND SPRING HEADED

'Christingle' (AGM) – Late. RHS describes it as having "Deep white curds of excellent quality."

**'Galleon' (AGM)* – Early mid-season. Medium to large curds; good leaf protection.

'Jerome' F1 (AGM) – Early mid-season. Very good quality.

MINI VARIETIES

**'Red Lion' F1 (AGM)* – Small summer variety with purple heads.

GW recommends *'Candid Charm'*, *'Clarke'*, *'Idol'*, **'Graffiti'* (purple curd) and *'Panther'* (green curd).

'Walcherin Winter Armado May' (AGM) and *'Walcherin Winter 3 – Thanet' (AGM)* – Both hardy.

Seed to harvest: summer and autumn cauliflowers 16 weeks; late winter and early spring varieties 40 weeks; mini varieties 15 weeks.

TURNIP, SWEDE AND KOHLRABI

Turnips and swedes are grown primarily for their roots, though the leaves can be eaten as winter greens. Kohlrabi is grown for its swollen stem (but not for the leaves). All are quite similar in taste. If you wanted to grow all three, I would suggest young turnips for spring, early summer and autumn; swedes (the hardiest of the three and the best for storing) for winter use; and kohlrabi for the height of summer.

TURNIP

Brassica rapa Rapifera Group

Turnips have fed both man and beast for many centuries. Before the advent of the potato, they formed an important part of the staple diet of the European poor, as well as being used as cattle fodder. Culpeper mentions them in passing as being "so well known as to need no description." The young spring turnip is quite different from the coarser mature version and can be a true delicacy.

Turnips were bred from the wild bitter ones that grow near streams and open land across Europe and Asia. The secret of successful turnip growing is speed. They are good for intercropping and undercropping.

Turnip types

Turnips come in flat, cylindrical and round varieties. The flesh is white or pale yellow and the skins may be white, yellow, pinkish or even black.

Cooking

Young turnips are good boiled and dressed with butter or olive oil and herbs. They can be served mashed, roasted, glazed or thinly sliced and stir-fried with greens. In the Middle East, long turnips are cooked and eaten cold with spicy dressing. They can be used in place of swedes for the traditional "neeps" (tur-neep) on New Year's eve. The tops can be used as spring greens. Turnips are a good source of vitamins A and C, as well as calcium.

Soil and situation

Turnips need a cool and moist situation; they can take a little shade. They like soil high in organic matter, and the pH no lower than 6.8. Prepare the bed in autumn by adding lime if necessary and firming well.

Sowing and planting

Earlies are sown in small batches every few weeks in March or April under cloches. Turnips don't like being transplanted, so sow them in situ. Sprinkle on a general fertilizer before sowing thinly in drills 2.5 cm (1 in) deep, in rows 23 cm (9 in) apart. Thin out to 10 cm (4 in) apart.

If you wish to carry on in May, find a shady spot. The maincrop ones are sown in July or August for eating from October onwards. Thin these to 20 cm (8 in) apart. A further sowing of turnips in autumn will give you spring greens in March.

Cultivation

Keep weed free and consistently moist. If turnips get too dry, they may bolt or the roots may become woody and unappetizing. If soaked after being dry, the roots may split.

Harvesting

Early turnips don't store; pull them up when they are the size of a golf ball; maincrop ones the size of a tennis ball before they coarsen. Twist off the leaves before digging. Store in a cool shed between layers of sawdust or sand.

Problems

Wireworm and cutworm.

Recommended varieties

EARLY

**'Ivory'* and **'Tokyo Cross' (AGM)* – Both are early, sweet-tasting varieties with a fine flavour.

**'Snowball'* – Introduced in 1869. GW recommends it as reliable.

MAINCROP

**'Golden Ball'* – A fast grower with yellow skin and flesh.

'Market Express' (AGM) – Late hardy variety, good flavour when eaten young.

**'Veitch's Red Globe'* – Introduced in 1882; red.

Seed to harvest: 6–12 weeks.

SWEDE

Brassica napus Napobrassica Group

The swede, a turnip cross that arrived from Sweden in the 18th century, is a larger and hardier species. Slow growing, it can stay in the ground in the coldest winters until late December before being lifted and stored.

Swede types

The tops of swedes that show above ground usually have a purple tinge, though there are green-topped and white-topped varieties. The flesh and skin are yellow. Earlies are for Christmas and lates for the New Year. New breeding has brought in resistance to club root and powdery mildew.

Cooking

Swedes are the classic ingredient for the bashed neeps on Burn's night. They are sweeter than turnips and have a lesser tendency to become watery when cooked. The roots are usually chopped and cooked slowly, then mashed in the same way as potatoes. The leaves can be eaten as winter greens.

Soil and situation

Prepare the bed in a sunny place, preferably on ground which was manured for a previous crop. Swedes like soil of pH7 or above.

Sowing and planting

Sow seed under fleece in March or in April to May, and thin out when the seedlings are 2.5 cm (1 in) high. Space them 23 cm (9 in) apart, in rows 38 cm (15 in) apart.

Cultivation

Water regularly and consistently. Drought makes the roots go woody, while too much water will affect the flavour. If drenched after drying out, they may split.

Harvesting

Lift swedes as you want them from autumn until Christmas. Leave a few for winter greens, then clear the whole of the remaining crop. To store, cut the leaves almost to the neck and keep in a cool, dry place. If you want to sweeten the winter greens, blanch them by packing in boxes and excluding the light in the same way as chicory.

Problems

Seedlings are prone to cabbage root fly and mealy cabbage aphid. Clubroot, mildews and weevils can affect the roots as they grow.

Recommended varieties

★'Best of All' – An old favourite, which is hardy and reliable.

'Joan' – Good for early sowing. HDRA describes it as having moderate resistance.

★'Marian' – Resistance to club root and mildew, and a good flavour.

'Ruby' and *'Invitation'* – These are both purple varieties known for their flavour.

Seed to harvest: 20–26 weeks.

KOHLRABI

Brassica oleracea Gongiloides Group

Pliny described kohlrabi as "the Corinthian turnip". Its German name translates as cabbage turnip, accurately describing the way it tastes. Perhaps there is something of the water chestnut about it too. Unlike swedes and turnips, it is not the root but the swollen stem that is eaten. Kohlrabi is reminiscent of a space craft in shape – a good talking point as it is not commonly grown here. As it has a spreading root system, it copes better with dry spells than turnips or swedes. It grows fast and, as an extra bonus, it rarely falls prey to cabbage diseases.

Kohlrabi types

Kohlrabi comes with pale green, white or purple skin. The purple types are slower but hardier and are usually planted later for autumn eating.

Cooking

It can be steamed as a summer substitute for turnips. When young it is good grated in salads. The leaves are not edible.

Soil and situation

Kohlrabi enjoys classic firm brassica soil, enriched the autumn before and limed where necessary to achieve a pH of 6–7. Choose a sunny site.

Sowing and planting

Kohlrabi can be grown under cloches from February onwards in mild areas, though it's safer to wait a little as a cold snap can make it bolt. The temperature needs to be over 10°C (50°F) for successful germination. For continuity, sow a few seeds every couple of weeks from spring onwards. Station sow them in threes, 2 cm (3/4 in) deep and about 23 cm (9 in) apart. Space rows 30 cm (12 in) apart. Thin early.

Start with the white varieties and move onto the purple ones. If you want winter crops, carry on until September.

Cultivation

Keep weeded and well watered in dry spells through summer.

Harvesting

Dig up when the stems are the size of a golf or tennis ball. They will carry on growing but coarsen as they turn into footballs! Cut off the roots and trim back the leaves. In mild areas they can be left in the ground in early winter. Eat shortly after harvesting.

Mini kohlrabi

Sow as above and thin to a planting distance of 2.5 cm (1 in) apart. Harvest when they are the size of a ping-pong ball.

 see pages 30–55

Brussels sprouts 'Igor'

Brussels sprouts 'Peter Gynt'

Cabbage 'Duncan'

Cabbage 'Offenham'

Cabbage 'Wivoy'

Cabbage 'Greyhound'

Cabbage 'Hispi'

Cabbage 'Stonehead'

Cabbage 'Celtic'

Cabbage 'Holly'

Cabbage 'Flagship'

Broccoli 'Bordeaux'

Broccoli 'Claret'

Broccoli 'Late Purple Sprouting'

Broccoli 'Red Arrow'

Cauliflower 'Plana'

Cauliflower 'Esmeraldo'

Cauliflower 'Marmalade'

Cauliflower 'Minaret'

Cauliflower 'Violet Queen'

Cauliflower 'Red Lion'

Cauliflower 'Galleon'

Cauliflower 'Graffiti'

Turnip 'Ivory'

Turnip 'Tokyo Cross'

Turnip 'Golden Ball'

Turnip 'Snowball'

Turnip 'Veitch's Red Globe'

Swede 'Best of All'

Swede 'Marian'

Kohlrabi 'Quickstar'

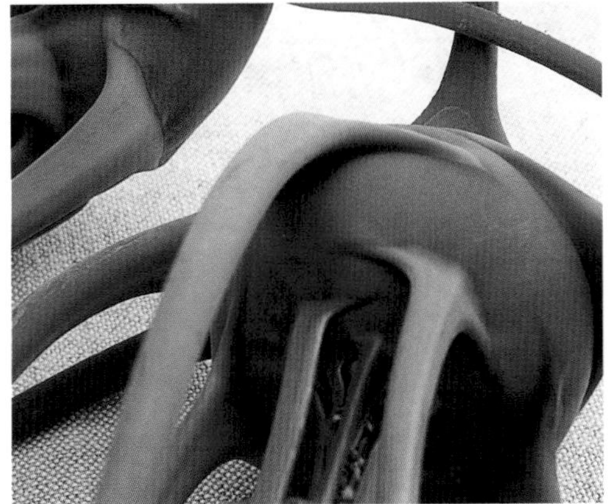
Kohlrabi 'Purple Vienna'

Kale 'Redbor'

Kale 'Red Russian'

Kale 'Reflex'

Chinese cabbage 'Early Jade Pagoda'

Chinese cabbage 'Ruffles'

Mizuna greens

Mibuna greens

 see pages 55–62

Carrot 'Amsterdam Forcing'

Carrot 'Bangor'

Carrot 'Comet'

Carrot 'Flyaway'

Carrot 'Sytan'

Carrot 'St Valery'

Carrot 'Yellowstone'

Carrot 'Purple Dragon'

Carrot 'Parmex'

Carrot 'Nantes'

Parsnip 'Javelin'

Parsnip 'Tender and True'

Parsnip 'Gladiator'

Parsnip 'Avon Resistor'

Jerusalem artichoke 'Fuseau'

Salsify 'Giant'

Sweet potato 'Beauregard'

Burdock

Problems

None.

Recommended varieties

'Azur Star' – A striking blue-purple variety.

'Kongo' F1 (AGM) – A large, green, fast-growing variety of exceptional quality. Sweet and moist, with high yield and attractive short leaves.

**'Quickstar' F1 (AGM)* – A fast-growing green type. Early and uniform.

'Purple Danube' (AGM) – Purple and sweet-tasting.

'White Danube' – Juicy white variety.

'White Vienna' and **'Purple Vienna'* – Traditional varieties.

Seed to harvest: white varieties 8–12 weeks; purple varieties 12–16 weeks.

KALE

(BORECOLE)

Brassica oleracea Acephala Group

It is said that kale was probably the first cultivated cabbage. Kales are very hardy and useful for the hungry gap. They are easy and fun to grow as some of the ornamental varieties are very striking and full of personality.

Kale types

There are four basic types. The Siberian kales, the collards (popular in the southern states of America), the Scotch kale (or borecole), and the ornamental kales. The ornamentals have real character. 'Chou Palmier' (the palm tree cabbage) looks like a cartoon of a palm tree. 'Chou Cavalier' (the giant Jersey kale) has a sturdy stem that is made into walking sticks beneath a topknot of leaves. There are types like 'Coral Queen' with handsome serrated green leaves and coral-coloured veins. New cultivars have been bred for sweetness.

Cooking

Use as greens. The flavour is said to improve after a frost.

Soil and situation

Can cope with poor soil. They like the same conditions as all brassicas. Choose a sunny site.

Sowing and planting

Sow in late spring for autumn and winter crops. Kales are usually sown in situ about 1 cm (1/2 in) deep. Thin to a final spacing of about 45 cm (18 in) apart for the smaller ones and up to twice that for the tallest ones.

Cultivation

Keep watered until established. Don't over water or over feed towards the end of summer to avoid a late flush of growth which could get cut back by the frosts. Earth up the taller types in autumn to avoid wind rock. Feed in early spring to encourage side shoots to grow.

Harvesting

Constantly cut off young leaves so that the plant doesn't have a chance to coarsen.

Problems

The usual brassica pests and diseases.

Recommended varieties

'Afro' (AGM) – Mid-green, very curly leaves. Compact. Stands well in winter.

'Jersey Kale' – The walking stick kale.

'Ragged Jack' – Pink in the leaves and mid ribs.

**'Redbor' F1 (AGM)* – Tall with purple-green leaves. Good for the hungry gap.

**'Red Russian' (AGM)* – Frilly green-red leaves.

**'Reflex' F1 (AGM)* – Curly blue-green leaves. Sweet tasting.

'True Siberian' – Harvests well all winter. Blueish frilly leaves.

Seed to harvest: 4–10 months.

TEXCEL GREENS

(ABYSSINIAN CABBAGE)

Brassica carinata

Texcel greens were developed

in 1957 from greens found in Ethiopia. The leaves are shiny and taste of spinach and cabbage combined, with a mustardy, garlic tang. As they grow at great speed, maturing in six weeks, they are a first-class choice for intercropping. They are widely used in agriculture as a catch crop and to provide game cover in winter. They do best in the cool of spring and autumn.

Texcel greens types

No cultivars as yet.

Cooking

The young leaves can be used as cut-and-come-again for salad mixtures and stir-fries. The older leaves can be cooked like spinach.

Soil and situation

Typical brassica soil. An open position with light shade.

Sowing and planting

Broadcast seed in situ every three weeks from early spring. A mid-autumn sowing under cover will provide winter crops. For small plants, thin to 5 cm (2 in) apart in rows 30 cm (12 in) apart. For large plants, thin to 30 cm (12 in) apart. For a cut-and-come-again crop, thin to 2.5 cm (1 in) apart.

Cultivation

Weed between plants. Water in dry periods.

Harvesting

Keep picking young outer leaves, as they will coarsen quickly.

Problems

Generally healthy, though they may succumb to brassica diseases. Flea beetle in hot weather.

Recommended varieties

None.

ORIENTAL BRASSICAS

The oriental brassicas that are becoming available in the West offer a whole new range to the repertoire of the gardener. They are fast-growing, naturally healthy plants, excellent for cut-and-come again. Most can be harvested at any time through their growing period for tasty additions to salads or stir-fries.

Some seed companies produce Oriental saladini – stir-fry and spicy seed mixtures, ideal for experimenting. Seeds can be sown in succession, outside in the summer and under cover in spring, autumn and early winter.

The flowering shoots of nearly all the oriental brassicas are a delicious part of their double act. Apart from the named varieties there are many sports grown primarily for their flowering shoots. Some of these, Choy sum ('Choy' is Cantonese for vegetable, 'sum' for flowering stem), Hon Tsai Tai (a purple-flowering version) and Flowering Pak Choi (bred specially for its flowering stems) can be found in British seed catalogues.

There are one or two points to watch when growing to maturity. Most are likely to bolt in the UK if they go short of water in the heat of summer and in spring if there are sudden drops of temperature. They are also sensitive to day length. The safest bet is to sow in late summer for autumn eating or to overwinter them under cover for spring.

Oriental brassicas need constant moisture, fertile soil and warm temperatures until they are established. Most have shallow roots that must not be allowed to dry out. New bolt-resistant varieties and F1 hybrids are being introduced from Japan all the time, so watch the catalogues for developments.

CHINESE BROCCOLI

(CHINESE KALE, GAI LAN)

Brassica rapa var. *alboglabra*

Chinese broccoli looks like a scaled down calabrese, growing to about half the height, 45 cm (18 in). The young flowering stems in bud are a true delicacy and crop over a long period. Chinese broccoli is probably the easiest of the oriental brassicas. It can take some heat and even mild frost.

Chinese broccoli types

Not many.

Cooking

The shoots are lovely in stir-fries with ginger and garlic. Any tough stems can be peeled and parboiled first. Chinese broccoli is high in vitamin C, beta-carotene and iron.

Soil and situation

Standard soil for brassicas, in a sunny site.

Sowing and planting

Though you can sow any time through summer, midsummer for autumn eating is the best time. Sow in situ 2 cm (3/4 in) deep and thin to 30 cm (12 in) apart. If you want to harvest the whole plants when young, space 10 cm (4 in) apart. An early autumn sowing under cover should bring a spring crop unless the winter is severe.

Cultivation

Keep well watered.

Harvesting

Lift the whole plant when young. Harvest the top shoot of full-sized plants to encourage side shoots to form.

Problems

Usually trouble-free.

Recommended varieties

There is a limited selection.

'Green Lance' F1 – Vigorous and fast maturing.

'Tenderstem' – A cross with calabrese.

'White Flowered' – Tall variety with blue-green leaves and white flowers.

Seed to harvest: 10 weeks to maturity, or six to harvest the whole plant young.

CHINESE CABBAGE

(CELERY CABBAGE, PEKING CABBAGE, PE TSAI, CHINESE LEAVES)

Brassica rapa var. *pekinensis*

Chinese headed cabbage is the familiar crisp vegetable that looks like a tightly packed pale green Cos lettuce. Recorded in ancient China, it probably came to Europe in the 18th century with the missionaries. It has never caught on as an amateur crop, being difficult to grow well in the European climate. A better bet are the loose-headed varieties of Chinese cabbage which will give three or four harvests of cut-and-come-again leaves.

Chinese cabbage types

There are cylindrical, barrel-shaped and loose-headed types. The loose-headed types are the easiest of the three to grow and the least likely to bolt. Some are ornamental and have frilly leaves. The Japanese F1 hybrids with built-in disease and bolt resistance are the safest choice.

Cooking

Headed cabbage is used in stir-fries or sliced as a salad green in the West. Loose-headed types can be used as ornamental salad leaves, depending on variety. Chinese cabbage contains moderate amounts of vitamin C and folic acid.

Soil and situation

Top-quality brassica soil is essential for success. Choose an open situation in sun, part shade in summer.

Sowing and planting

For headed Chinese cabbage,

the safest time to sow is in July to crop at the end of September or in October. Sow outside in situ 1.5 cm (1/2 in) deep, 30 cm (12 in) apart.

If you are growing loose-leaf varieties for cut-and-come-again, try a few seeds every couple of weeks, starting in late spring using fleece or cloches for protection.

Cultivation

Water regularly and keep soil weed free. If you are attempting headed cabbage, tie the leaves together with soft twine in late summer to blanch them.

Harvesting

When the heads are solid and before the frosts, cut them off at the stalk. They will stay fresh for a week or two in a cool frost-free shed or in the fridge.

Problems

The headed Chinese cabbages seem to attract all cabbage pests and diseases. They also suffer from bolting.

Recommended varieties

**'Early Jade Pagoda' F1 (AGM)* – Cylinder-shaped with crisp, dark green leaves. Resistant to bolting.
'Mariko' F1 (AGM) – Barrel-shaped with attractive outer foliage and a blanched heart.
'One Kilo SB' (AGM) – Solid barrel-shaped heads. Slow to bolt.
**'Ruffles'* – Another loose-headed type, green with a white heart.
'Santo Serrated Leaved' – Pretty leaves for salads when used as cut-and-come-again crop.

Seed to harvest: headed Chinese cabbage 8–10 weeks; cut-and-come-again 3 weeks.

KOMATSUNA

(MUSTARD SPINACH)
Brassica rapa var. *perviridis*
Komatsuna is a five-star plant, ready for harvesting like spinach in four weeks, and reaching its full potential in less than two months. A cross between mustard greens and turnip, it can be eaten at any stage and successfully sown from spring, right through summer. A late-summer sowing will make for a winter or spring crop under cover. When it does go over you can eat the tasty flowering shoots. If left to mature it grows up to 50 cm (20 in) in both width and height. The leaves are big and glossy and grow back when harvested, with vigour.

Komatsuna types

Crosses with pak choi and turnip greens are said to be in the pipeline.

Cooking

It can be steamed, stir-fried, or mixed with other greens like spinach. Some people find the taste too cabbagey to use in salads.

Soil and situation

They like classic brassica soil and an open site.

Sowing and planting

Mature crops Sow in July for autumn and early winter harvesting, and late autumn for winter use. The thinnings can be used for salads and stir-fries. Sow in drills about 45 cm (18 in) apart.
Cut-and-come-again Make both the first and last sowings of the season under cloches or fleece in early spring and autumn. From late spring onwards, sow successionally where you want them to grow. Thin to 2.5 cm (1 in) apart.

Cultivation

Keep weed free and well watered.

Harvesting

When harvesting for cut-and-come-again, always leave about 2.5 cm (1 in) of the plant to regrow. The mature plants can be cut whole or leaves can be taken over quite a long period. When the plant flowers, keep picking them for more flowering greens.

Recommended varieties

'Big Top' F1 – Greens plus turnip-like roots.
'Tendergreen' F1 – Fast grower; first pickings in 20 days.

Seed to harvest: 7 weeks in summer; 10 weeks in winter; cut-and-come-again 4 weeks.

MIZUNA AND MIBUNA GREENS

(MIZUNA MUSTARD, KIYONA)
Brassica rapa var. *nipposinica* and *Brassica rapa*
The name mizuna comes from the Japanese word for 'juicy or water vegetable' – a characteristic of the leaf stalks of these plants. They are a very practical proposition for Europe, being both hardy and heat-tolerant.

Mibuna greens are a fairly new introduction. They get their name from Mibu, Kyoto, where they have been cultivated for centuries. They closely resemble mizuna, although they have different-shaped leaves – long and elegant. They have a stronger taste and are less resilient.

Mizuna and mibuna types

Not many in the UK.

Cooking

The young leaves are an ornamental addition to the salad bowl all year round. The older leaves are peppery and can be cooked as greens. The leaf stalks take longer to cook than the leaves and are usually done separately.

Soil and situation

They like classic brassica soil, though mizuna and mibuna greens are not as fussy as most. They enjoy an open site in sun, though in the heat of summer light shade is a better choice. Both are good for intercropping between taller vegetables – Brussels sprouts for example.

Sowing and planting

Mizuna In spring when the weather is set fair, sow a few seeds under fleece or cloches in situ. Carry on through the summer and into autumn for seedling crops. Mizuna doesn't grow as well in the heat of summer so take a break from it. For mature-headed plants in winter and early spring, sow under protective cover in autumn.
Mibuna Don't sow until late summer for an autumn crop of cut-and-come-again. Sow in autumn under cover for winter pickings.

Space cut-and-come-again crops 5 cm (2 in) apart, small plants 10 cm (4 in) apart, and medium plants double that spacing. For full-headed plants, space at 30–45 cm (12–18 in) intervals.

Cultivation

Very little needed. Keep weed free and watered.

Harvesting

You can almost see mizuna and mibuna greens grow. The first leaves can be harvested within three weeks. To keep up a supply of young leaves over many months, it's important to harvest almost on a daily basis.

Problems

Few, though watch for slugs and flea beetle.

Recommended varieties

*MIZUNA GREENS
'Mizuna Greens' – Generally sold as mizuna greens.
'Tokyo Beau' – Has good cold resistance.

*MIBUNA GREENS
'Green Spray' F1 – An early variety.

Seed to harvest: cut-and-come-again 2–3 weeks; mature plants 8–10 weeks.

MUSTARD GREENS

(KAI TSOI, SARSON KA SAAG)
Brassica juncea
Mustard greens are a huge and

diverse group of leafy plants from China. They come in a huge range of shapes and colours – green, red or purple, with jagged or smooth edges. They are highly nutritious and get hotter as they run to seed.

Mustard green types

The lower leaves of mustard greens are up to 60 cm (24 in) long. They are a hardy winter crop. The red-leaved varieties, including 'Red Giant', are sometimes grown as seedling crops through summer and are excellent for cut-and-come-again. Giant mustards are cultivated in the same way as Chinese cabbage. Start them off in July in cold areas (August or September in warmer ones) to grow on through winter. Common mustards are more difficult, tending to bolt in spring.

Green-in-snow types are fast growing and hardy. Being quite pungent, they are distasteful to pests; and they are resistant to disease. Curly-leaved mustards are not unlike curly kale in appearance. These are robust plants for winter consumption.

Cooking

Young leaves can be eaten raw. The stalks are removed and the leaves of mature specimens are cut into ribbons before being cooked like any other spring green. The flowering shoots are also eaten.

Soil and situation

Provide normal brassica soil. Mustards are not too fussy.

Sowing and planting

Mustard seed is very fine. Sow sparingly on the surface of the soil in situ and sieve over a thin layer of compost. Varieties vary in their needs. As a rule of thumb, sow mid to late summer for winter eating, and from spring onwards for seedling crops.

Space seedling crops 15 cm (6 in) apart and gauge upwards according to the ultimate size of the mature plants. The giant mustards should be thinned to 60 cm (24 in) apart.

Cultivation

Keep well watered.

Harvesting

Pick cut-and-come-again according to your needs. Small plants are often harvested whole when about 15 cm (6 in) high. Large plants can be treated in either way. By spring most will run to seed.

Problems

Rare. Can be affected by flea beetle, cabbage root fly, aphids and slugs.

Recommended varieties

'Art Green' – A curly-leaved type. Fairly versatile and comparatively good in both high and low temperatures.
'Green in the Snow' – Has jagged leaves.
'Red Giant' – Large plant with wrinkly red leaves.

Seed to harvest: 6–8 weeks to maturity.

PAK CHOI

(CHINESE CELERY MUSTARD, CHINESE WHITE CABBAGE, MUSTARD CABBAGE, BOK CHOI, JAPANESE WHITE CELERY MUSTARD)
Brassica rapa var. *chinensis*

With a little care and attention to avoid bolting, pak choi is a practical proposition to grow in the UK. Every bit of the plant can be eaten and it can be harvested at any stage of its development. It grows fast, maturing in about six weeks, making it a good choice for intercropping.

Pak choi types

There are many different types, though they are not always clearly defined or widely available. There are three main types. Chinese white is a sturdy plant about 30 cm (12 in) tall, with thick leaves and very white midriffs.

Soup spoon types are more refined, with cupped leaves and overlapping leaf stalks, reminiscent of Chinese soup spoons. These are a little taller at 45 cm (18 in) and more elegant. Canton types are more squat and considered to have the best flavour.

Cooking

Use in salads, stir-fries, as greens or garnishes. The flowering shoots are good raw or steamed.

Soil and situation

Carefully prepared brassica soil, in an open situation with semi-shade in summer.

Sowing and planting

Sow seedling crops outside under cover in spring. Seeds can be broadcast or sown in drills. Sow in July for mature plants in autumn. In Autumn (six weeks before the frosts) sow hardiest types to grow under cover for winter eating.

Thin according to the ultimate size you want and the variety. A medium-sized type would be spaced about 20 cm (8 in) apart.

Cultivation

Water well.

Harvesting

Pak choi should be harvested in the peak of good health. It doesn't store for more than a few days in the fridge.

Problems

Can be subject to all the brassica problems.

Recommended varieties

'Autumn Poem' F1 – Designed to produce tender flowering shoots.

'Joi Choi' F1 – A vigorous hybrid bred for our climate with frost and bolt resistance. With green leaves and white mid ribs, it resembles Swiss chard in appearance.

'Mei Quing Choi' F1 – This has been bred to produce small tender heads for close planting after midsummer.

'Pak Choi – Canton Dwarf' – A mini variety with white stems and green leaves. Tolerant of heat.

Seed to harvest: 10 weeks to maturity; 2 weeks for cut-and-come-again.

ROOTS

The root vegetables come from many different families but need much the same care and conditions. As all produce their crops below ground, they need stone-free, well-cultivated soil to grow long straight roots. If you have heavy clay soil, you can get round it by making a V-shaped trench or individual cornet-shaped holes and filling it, or them, with proprietary compost or suitable sieved garden soil. The alternative is raised beds or growing in barrels.

They do best on soils with plenty of organic matter, but not recently manured as this can make them fork. Root crops are generally sown from seed where they are to grow. They don't transplant well or respond well to checking. Aim to grow them fast and in ideal conditions.

Watering is a fine balance. Root crops need to be encouraged to search for moisture below ground. If they are overwatered they are inclined to put out more leaf than root. On the other hand, if they are allowed to dry out and then get soaked, they may fork or bolt.

Protect early crops against cold with fleece. Common root diseases are cankers and mildews. The carrot fly can devastate carrots and parsnips. Rotation is important to help keep down soil-borne pests, particularly eelworm.

CARROT

Daucus carota

The wild carrot, with its small, acrid, whitish root, is the familiar and pretty Queen Anne's Lace of our pastures and roadsides. Depictions of carrots in many colours – purple, white, yellow, red and green – can be seen in art going back as far as Ancient Egypt in 2000 BC. Unpleasant tasting, the wild carrot was not eaten as a vegetable by the Greeks and Romans, only used in medicine. The yellow carrot was recorded in Turkey in the 10th century, while both yellow and dark purple ones became quite widely cultivated in Europe.

It was not until the 16th century that the breakthrough orange carrot appeared. It would seem that it was bred by the Dutch horticulturalists to celebrate the House of Orange. By that time, carrots were considered an epicure plant. The courtesans of the Stuart court would wear the young foliage pinned to their hats like feathers. All cultivated carrots today are descendents of the four Dutch types bred in this era – Early Half Long, Late Half Long, Scarlet Horn and Long Orange. Since the time of Elizabeth I, carrots have been one of the most widely grown vegetable crops.

Carrot types

Today we have the full range – long, round or short, and yellow, white or purple as well as orange. There are different types for autumn and summer planting and for forcing. When buying seed, check the packet for the recommended season as new breeding makes the picture complicated. As a general guideline, use the Amsterdam Forcing types for the earliest crops. Carry on with these, as well as the fast-growing short Round and Nantes ones for spring. The Paris Market carrots have small, round or square roots and are the best bet for soil on the heavier side. These early types can also be sown in autumn under fleece or cloches for a spring crop. The Chantenay, Berlicum and Autumn King cultivars are large maincrop types for summer eating and winter storage.

Cooking

Carrots are widely eaten raw, or cooked in stews and soups. They can be used in sweet dishes – for carrot cake and charlottes. In India they make carrot halva to be eaten with cream. Carrots are a great source of anti-cancer beta-carotene which converts in our bodies to vitamin A.

Soil and situation

Carrots need light sandy loam and constant moisture. The site should be sheltered and sunny for earlies, and open and sunny for maincrop types.

Sowing and planting

Sow under a cold frame in February in mild areas (more safely in March) when the soil temperature has reached a minimum of 7°C (46°F). It is said that if you wait until the soil is a little warmer, the carrots will grow twice as fast. Sow in September with protection for the following spring.

Make drills 1 cm (1/2 in) deep. The seed is like fine dust, so scatter it carefully, a pinch at a time. If you find this tricky, mix the seed with some silver sand or try fluid sowing (see page 156). Sift a very thin covering of compost over the seed. Some gardeners protect the seeds further by covering them with a low ridge of soil, removing it after a week when seeds have germinated. Space rows 30 cm (12 in) apart.

Protect the seedlings from carrot fly with a low barrier of netting. Sow a few seeds outdoors every two weeks from April onwards and you'll have a succession of carrots until early winter. For winter storage, sow in May or June to harvest in October.

Cultivation

Thin the carrots to 10 cm (4 in) apart. Firm the soil well after thinning and weeding. Keep carrots on the dry side for sweetness of flavour but not so dry that they will fork. Maintain a low but constant level of moisture. Mulching between rows helps.

Harvesting

If you have light soil, carrots can be pulled up by hand, otherwise the soil may need to be watered before they are eased out with a fork. Try to make minimum disturbance and damage as the smell will alert the carrot fly. If you have light, sandy loam, maincrop carrots can be left in the ground until needed. Protect them with straw before the onset of frost. Otherwise lift the crop, twist off the foliage and store in a cool dry shed in boxes of sand.

Problems

Carrot fly.

Recommended varieties

**'Amsterdam Forcing'* – The traditional tried-and-tested first carrot of the year.

'Autumn King 2' (AGM) – Fine quality, cylindrical, deep orange roots. Stores well and can be left in the ground for some time without losing flavour.

**'Bangor' F1 (AGM)* – GW recommends it as the "best maincrop on most soil types." It has cylindrical roots and heavy yields, and stores well.

'Danvers' – Victorian carrot for successional summer crops.

'Early Nantes' – An old favourite for successional planting. **'Nantes 2'* follows for an early main crop.Virtually coreless. GW recommends it for early finger carrots.

**'Flyaway' (AGM)* – A stumpy carrot with sweet orange flesh. Some resistance or lack of attraction to carrot fly.

'Giganta' – A large Autumn King type. Good show carrots.

'Kingston' (AGM) – Autumn King type. A handsome carrot, long and pointed. Good for showing and autumn storage. RHS describes it as "one of the best for colour, taste, yield and uniformity."

'New Red Intermediate' – A fine carrot for exhibiting. Stores well.

'Panther' F1 (AGM) – A stumpy, quick-growing maincrop. Recommended for successional plantings.

'Parabel' (AGM) – Sweet round roots. Early French Market type. Good flavour.

'Resistafly' – As the name suggests, it has resistance to the carrot fly. Also recommended by GW.

'Sutton's Red Intermediate' – Large carrot and a favourite for showing.

**'Sytan' (AGM)* – Another breakthrough carrot with some built-in resistance to carrot fly. Good flavour. Sweet-tasting tapering roots.

Seed to harvest: 9 weeks for earlies, 20 weeks for maincrops.

PARSNIP

Pastinaca sativa

It is said that Emperor Tiberius held parsnips in such high esteem that he had special deliveries of them from the Rhine Valley. In 16th-century Germany, parsnips were another winter staple of the European poor before the introduction of the potato. Traditionally they were grown to be a large coarse vegetable. They would be given the best part of a year to mature and left until after the first frosts to sweeten. They were widely grown to feed livestock, particularly pigs. This has given them an unfairly low reputation. New developments in breeding have resulted in a more elegant product in the form of small nutty parsnips that are available all year round.

Parsnip types

Short and long, narrow and bulbous.

Cooking

Parsnips were traditionally used to sweeten cakes, to make jam and wine. In the USA they are glazed with fruit juice and sugar. The French use them in the *pot au feu* as a flavouring vegetable. They are good parboiled and roasted.

Soil and situation

Parsnips are less particular than carrots but the ideal is the same. Choose an open, sunny site.

Sowing and planting

Parsnips are notoriously slow and erratic to germinate. Buy seed fresh each year and delay sowing until late spring. Give them a head start with pre-germination, growing in biodegradable modules or by using the fluid sowing technique (see page 156).

If you are sowing directly outside (which they prefer if conditions are right), prepare the ground carefully and warm it for a week or so with polythene. Make drills 2.5 cm (1 in) deep with 30 cm (12 in) between rows. Sow three seeds at each station, 12.5 cm (5 in) apart for medium-sized parsnips). Germination can take up to a month. When the plants are about 5 cm (2 in) tall, thin to the strongest.

Cultivation

Control weeds. Provide constant moisture.

Harvesting

Parsnips are ready when the leaves start to droop in autumn. They are generally left in the ground, becoming sweeter after the frosts. The leaves die right down so mark the spot. Lift crops by the end of February as they will start to grow again. Consider leaving a few as they will produce tall flowers that will draw in beneficial insects.

Problems

Carrot fly, parsnip canker.

Recommended varieties

'Cobham Improved Marrow' (AGM) – Canker resistant, a popular stubby, wedge type with flavour.

**'Gladiator' FI (AGM)* – High resistance to canker, vigorous and with good yields. Smooth skin.

**'Javelin' F1 (AGM)* – Wedge-shaped roots. Good yields. Canker resistance.

'Lancer' – A good variety for mini carrots. Plant close together.

**'Tender and True' (AGM)* – An old variety. Nearly coreless, sweet-tasting and canker resistant. Tapering roots for deep soil and exhibition. Sizes are variable.

'White Gem' – This variety has broad shoulders and white skin and is resistant to canker. Heavy cropper. Sweet flavour and not too fussy about soil.

JERUSALEM ARTICHOKE

Helianthus tuberosus

The Jerusalem artichoke has nothing to do the artichoke, though it has a similar taste. Related to the sunflower, it's known as a 'sunchoke' in the USA. It grows wild in good marshy soil from Georgia to Nova Scotia. The French explorer Samuel de Champion sent some back to France in 1605. Some 30 years later they reached Italy where they were called *girasole* (sunflowers). Possibly Jerusalem is a corruption of this.

These plants make a leafy screen up to 3 m (10 ft) high – excellent for privacy in summer or as a wind break, also as cover for wildlife.

Jerusalem artichoke types

The traditional Jerusalem artichoke has a knobbly difficult-to-peel root. New cultivars include smoother tubers and some that don't need peeling. Buy plump healthy tubers, the size of an egg, from the greengrocer to plant in early spring.

Cooking

Jerusalem artichokes are delicious in soups. They make an excellent cooked vegetable and can be used as a substitute for water chestnuts in Chinese cookery.

Soil and situation

Any soil, in sun or shade.

Sowing and planting

Large tubers can be cut into sections with one or two eyes. Plant in early spring 10–15 cm (4–6 in) deep and 50 cm (20 in) apart. If you want more than one row, space them 75 cm (30 in) apart.

Cultivation

Stake and earth them up if you are on a windy site. Water in dry spells. If any sunflower blooms appear, remove them to improve the crop.

Harvesting

Cut them right down after frost has killed off the tops. Spread the tops over the patch for winter protection. Dig the tubers up as you want them through winter. Clear the ground in early spring, keeping a few to replant.

Problems

They can become rampant.

Recommended varieties

'Dwarf Sunray' – A variety which doesn't need peeling.
**'Fuseau'* – An old variety with long, smooth tubers.
'Gerrard' – Very smooth purple skins, round rubbers.
'Mules Rose' – Big white tubers. Vigorous.

Planting to harvest: around five months.

CHINESE ARTICHOKE

(KNOT ROOT, CHORGOI)
Stachys affinis

The Chinese artichoke, known in Europe since 1882, has sage-like leaves, pretty pink flowers and grows to about 45 cm (18 in) tall. The tubers, which are segmented like a caterpillar and grow in strings at the ends of the roots, are about 5 cm (2 in) long and as thick as a fountain pen. They taste like water chestnuts and artichokes and are said to be highly prized in China as they resemble white jade.

They are easy to grow though you need to plant quite a few for a decent crop. When the plants have flowered and the first frosts have blackened the leaves, dig them out as you want them, as they don't store for long. In spring take care to clear the ground thoroughly as they can take over. Keep a few for the next crop.

SALSIFY

Tragopogon porrifolius

Salsify, an annual meadow plant from the Mediterranean, has strappy leaves and pretty pink to purple flowers and can be grown as an ornamental. The white roots grow to 23 cm (9 in) long, have a faint taste of oyster, hence the common name of oyster plant or oyster vegetable. John Tradescant the Younger mentioned salsify in 1656.

Salsify is having something of a renaissance, as is scorzonera. Both belong to the daisy family, are hardy and are grown and cooked in the same way. Not commonly available in the shops, they make an interesting and easy crop for the allotment. You need to grow quite a few, however, to get much to eat.

Salsify types

There are some interesting cultivars.

Cooking

Can be cooked as a vegetable, made into soups or added to pies. A classic dish from the Périgord region in France is chicken and salsify pie, *tourtière aux salsifis*. Both the cooked roots and the leaves can be used in salads.

Soil and situation

Classic root soil. An open sunny situation.

Sowing and planting

Sow the seed while still fresh in situ in early to mid-spring. To get straight roots, make funnel-shaped holes 30 cm (12 in) deep and 23 cm (9 in) apart with an iron bar. Fill with compost or sieved sandy soil. Sow three seeds per station as germination can be erratic. Thin to one when the plantlets are about 5 cm (2 in) high. If this is done carefully, you may be able to transplant them.

Cultivation

Hand weed as the roots may bleed. Keep steadily watered to avoid forking.

Harvesting

The roots should be ready by autumn but can be left all winter. Lift with care as they are brittle. Use promptly before they shrivel. For tasty shoots and flower buds (chards) that can be enjoyed like asparagus in spring, cut the leaves down to about 2.5 cm (1 in) in autumn.

Problems

Unlikely.

Recommended varieties

**'Giant'* – A variety with long roots.
'Mammouth Sandwich Island' – Recommended by GW as best in trial.
'Mammouth White' – Recommended by HDRA.
'White Skinned' – A very hardy variety.

SCORZONERA

Scorzonera hispanica

Scorzonera, also known as 'black serpent' due to its skin colour and shape, originates from Spain. The main difference between it and salsify is in the colour of the skin and the fact that salsify is an annual and scorzonera a perennial. The roots are both longer and thinner than salsify, rather like a long cigar. They have a delicate nutty taste but are less sweet than most root vegetables. Treat as salsify. They are perennials so if roots are too small for harvesting in autumn, leave the plant to grow on.

Recommended varieties

'Habil' – This is possibly the best for flavour.
'Long Black Maxima' – Long tapering roots and a good flavour.
'Maxima' – Variety with long black roots.
'Russian Giant' – A long-rooted, black-skinned, very hardy variety. Widely recommended.

SWEET POTATO

Ipomoea batatas

The sweet potato, a South American native, is widely grown in tropical countries. Until recently no one would have considered growing it outside a hothouse in the UK. However, you can now buy sweet potato slips (rather like a seed potato), to give you a head start in our short summers. In truth, however, results are generally reported to be disappointing with undersized tubers. No doubt breeding will toughen it up for our climate in a few years, as it did with the ordinary potato.

Sweet potato types

Sweet potatoes come in white, pink, purple and orange but the orange types are the best.

Cooking

It can be used in the myriad ways that potatoes can be cooked. Americans enjoy it glazed with honey or maple syrup and fruit juice. In South

America it is a popular ingredient for puddings and pies as well. It is rich in vitamins A and C and beta-carotene.

Soil and situation

If you want to go ahead, aim for a little patch of Peru – rich sandy soil (pH5.5–6.5), dug over well with general fertilizer added.

Planting

Plant in May when the soil has reached at least 12°C (52°F). Warm it with black polythene or cloches before the tubers arrive. They can be stored for a few days with their roots in water. Plant them 15 cm (6 in) deep and 30 cm (12 in) apart with at least two leaf nodes buried under the soil. Grow under a floating mulch (see page 167).

Cultivation

Keep well watered. Give a high-potash feed every fortnight.

Harvesting

With luck they can be harvested in the same way as potatoes in September. Dry in the sun or shed for a few days. They will store for about a week at 15°C (59°F).

Problems

Leaf spot, sooty mould, black rot and brown rot. Under cover, red spider mite may appear.

Recommended varieties

**'Beauregard'* – An orange variety and the first to be supplied as a slip.
'Centennial' – Reputed to be vigorous and fast-growing, so suited to our climate.

HAMBURG PARSLEY

Petroselinum crispum var. *tuberosum*

Hamburg parsley has a root like a slender parsnip and tastes rather like celery. It can be cooked or grated raw into salads like celeriac. The leaf has a strong parsley taste – too strong for some. The leaves stay healthy and vibrant through winter, providing a useful addition for the kitchen.

Hamburg parsley types

Not widely available.

Cooking

Two vegetables in one, it is widely grown in Eastern Europe for soup. Hamburg parsley came into its own here during the Wars. The roots can be eaten roasted, boiled or mashed. The leaves can be used for flavouring stocks.

Soil and situation

The ideal is sandy loam (pH6.5) and an open sunny site, although anything will do.

Sowing

Work the soil to a fine tilth and warm it in early spring. Sow thinly in rows 1 cm (1/2 in) deep and 25 cm (10 in) apart, or station sow three seeds at intervals, thinning to the strongest when they are 5 cm (2 in) high. Germination is slow. You can also sow in midsummer for crops the following spring.

Cultivation

Keep well watered. Hoe carefully when young to avoid damage.

Harvesting

Ready by autumn, it can be left all winter and harvested fresh as wanted. Alternatively, the roots can be stored in a dry shed in sand, though they will gradually lose quality.

Problems

Generally trouble free. Erratic, slow germination. Parsnip canker.

Seed to harvest: 30 weeks.

*BURDOCK

(GOBO, NGAO PONG)
Arctium lappa

Burdock is a classic vegetable

in Japanese cuisine. The slim white taproots are only 2.5 cm (1 in) across but can grow ridiculously long – up to 1.2 m (4 ft). They are usually harvested when they reach about 30 cm (12 in) before they become impossible to dig up. Burdock has a crisp texture and is described as tasting somewhere between scorzonera, parsnip and Jerusalem artichoke.

The wild form is native to Europe, Asia and America. It is a common weed found in damp waste ground and road verges.

The entire plant grows to about 1.2 m (4 ft) with bare branched leaf stalks topped by coarse heart-shaped leaves up to 30 cm (12 in) across. In late summer the purple thistle flowers turn into burs that will stick to any passing creature.

Burdock types

There are brown- and white-skinned varieties.

Cooking

In Japan the root is eaten raw when young, pickled or served with sweet sauce. It can be cooked in the same ways as any other root vegetable. The coarse heart-shaped leaves can also be cooked and eaten when young. Young roots and shoots can be pulled up and cooked in the same way as young beets.

Soil and situation

They need classic root soil. Dig deeply to allow the root to grow down 30 cm (12 in). Plant in sun or part-shade.

Sowing and planting

Either sow in early spring for autumn harvesting or in late autumn for the following summer. Joy Larkcom recommends the following method. For overwintering crops, wait until late autumn. Soak the seeds overnight, sow them in a seed tray, keep at 20–25°C (68–77°F) and leave them uncovered. Transplant outside before they develop a tap root. They will pick up and grow fast in spring. To encourage straight roots, plant closely together, about 10–22 cm (4–9 in) apart.

Cultivation

Weeding and watering.

Harvesting

Not the easiest task. Dig deeply and tug. They will keep for a few weeks in the fridge.

Problems

Bolting.

THE ONION FAMILY

Onions have been cultivated for thousands of years. In the *Book of Numbers* the children of Israel complained to Moses, saying they missed the onions, garlic and leeks of Egypt.

There is nothing too difficult about growing onions as long as you get the soil right and the summer is fair. The soil should be open, free draining and fertile. Onions dislike acid soil so liming may be necessary to bring the pH to 7 or more. Incorporate rotted compost the autumn before as freshly manured soil can encourage bulb rots.

It is important to put them into a four-year rotation plan to help to prevent soil-borne pests and disease. Onions need sunshine so they are easier to grow in warmer parts of the country. All need an open site with plenty of air circulation to avoid downy mildew. The aim is to grow them dry and hard. Once they've established, only water them if they show signs of wilting.

Problems with bulb onions can be mildews and rots, particularly onion white rot and onion neck rot. Others are onion eelworm, onion fly and bolting. Sparrows like to pull the newly planted onion right out of the soil. If this happens all is not lost. Dig them out carefully, replant and protect with fleece, netting or thread stretched across the rows between short stakes.

BULB ONIONS AND SHALLOTS

Allium cepa

The bulb onion was first cultivated in western Asia and spread to all the ancient civilizations. In the 1st century AD, bulb onions of many differing types were recorded – yellow, red and white, strong tasting and mild, long and round. They were widely grown and eaten in the Mediaeval era in Europe, and have always been valued for their long shelf life. Some varieties will store right through winter.

Onion and shallot types

There are globe, elongated (flattened) globe, and spindle types. Skin colours are white, through yellow and brown, to purple. There are different types for different seasons. For mini onions, use any type and plant close together.

Maincrops The maincrop types are sown in spring (seed or sets) and harvested in late summer or autumn. They can be stored through winter.

Japanese onions *Allium fistulosum*, (seed or sets) can be sown or planted in late summer or early autumn to overwinter and provide onions the following summer. These don't store as well but will keep you supplied until the maincrop ones are ready in autumn. Breeding has also brought in autumn sets of other globe onions. These are treated in the same way for harvesting in June.

Shallots These small bunching onions can take more heat and more cold than the maincrop type and store extremely well. Shallots are grown from sets, as seed will only produce a single bulb rather than a bunch. They are planted in winter (early or late depending on which area) and will be ready to eat between the end of the stored maincrop ones and the beginning of the Japanese and autumn types.

If you grow all types, you will have onions all year round.

Cooking

There is hardly a savoury dish that isn't enhanced by onions. In her celebrated *Vegetable Book,* Jane Grigson notes that "no civilization, no developed cuisine could do without onions. They can be boiled or baked and stuck with cloves or added to dishes by the score as a combination of lubricant and flavouring."

Soil and situation

See general notes on the onion family. If sowing seed, rake the soil to a fine tilth.

Sowing and planting

Buying sets of certified bulb onions is recommended for ease of cultivation, speed, and less tendency to disease. The heat-treated ones have the advantage of bolt-resistance. Growing from seed gives you a wider choice. Some say that the onions grown from seed store for longer than those grown from sets. It is important to get the sowing time right for seed, as onions are sensitive to day length and only start to form bulbs from midsummer onwards.

Growing from sets Plant sets outside under cloches when the ground is workable in late spring. Heat-treated sets go in later. Plant them with the tips just showing. Space 15 cm (6 in) apart for average-sized onions, in rows around 30 cm (12 in) apart. Adjust spacings for larger or smaller ones.

Sowing seed In early spring, start the seed off indoors at 10–16°C (50–61°F). Keep them on the cool side when germinated, about 13°C (55°F). Onions grown from seed first send up a crook, a shoot that forms a loop. Don't try to free it because it has a purpose. It is drawing nourishment from the seed in the ground and will straighten itself. When onions reach the crook stage they are ready to be pricked out. Harden off and transplant out when they have two true leaves and the soil is warm enough.

For late spring sowings, warm the ground if necessary. Sow seed outside under cloches when the ground is workable. Follow the timings on the packets. Use the thinnings as spring onions.

Japanese onions The planting time for Japanese onions is critical. Too early and they may bolt in spring; too late and they won't survive the winter. You want them to be 15–20 cm (6–8 in) tall by the first frosts. If you are growing from seed, sow a few outside at two week intervals through August, earlier in the north and later in the south. Sets are sturdier and can be planted from September to November.

Cultivation

Keep well weeded and watered until established, then only water if they begin to wilt.

Harvesting and storing

Lift the bulbs when the foliage dies back. If the weather stays fair, leave them in the ground until the leaves are crackly dry. Once lifted, lay them out in the sun to dry off in a single layer. Alternatively, bring them under cover or lay them on sacks under a cold frame. Leave for a week or two until the skins are brown and papery. Don't remove the leaves until they are completely dry or the onions won't store well. They can be hung in nets, made into plaits or laid out on shallow trays in a single layer.

You can also harvest onions when still green, or just ripened in late summer for immediate eating.

Problems

Prone to onion pests and diseases.

Recommended varieties

GLOBE ONIONS FROM SETS IN SPRING

'Centurion' F1 (AGM) – A heavy cropper, early maturing and a good storer. Flattened globe shape; straw-coloured. Recommended by GW.

'Hercules' F1 (AGM) – Elongated shape and resistant to bolting.

**'Sturon' (AGM)* – Big globe-shaped, yellow-brown onions. Resistant to bolting, excellent flavour. Stores until the following spring.

'Turbo' (AGM) – A globe type with golden skin. Slow to bolt, with a good yield. Recommended by GW.

Sets to harvest: 18–20 weeks when planted in spring.

GLOBE ONIONS FROM SEED IN SPRING

**'Ailsa Craig'* – An old-fashioned heavy cropper. The winner of many shows.

'Rayolle de Cevennes' – A big onion with yellow skin, mild and sweet from Cevennes in France.

**'Red Baron'* – Red-skinned onion with concentric red and white rings inside. Recommended by GW.

**'Rijnsburger 5 Balstora' (AGM)* – A pale yellow globe. Good keeper.

Seed to harvest: 20–24 weeks.

OVERWINTERED BULB ONION GROWN FROM SEED

**'Buffalo' F1 (AGM)* – Good yielder of flattened globes.

THE ONION FAMILY see pages 62–76

Onion 'Sturon'

Onion 'Red Baron'

Onion 'Rijnsburger'

Onion 'Ailsa Craig'

Onion 'Senshyi Yellow'

Onion 'Buffalo'

Shallot 'Sante'

Pickling onion 'Paris Silverskin'

Pickling onion 'Purplette'

Spring onion 'White Lisbon'

Japanese bunching onion 'Ishikura'

Japanese bunching onion 'Redmate'

Garlic 'Cristo'

Garlic 'Elephant'

Garlic 'Early Wight'

Leek 'Toledo'

Leek 'Jolant'

Leek 'King Richard'

BEETROOT, SPINACH, OTHER GREENS see pages 77–84

Beetroot 'Boltardy'

Beetroot 'Bulls Blood'

Beetroot 'Burpees Golden'

Beetroot 'Chioggia'

Beetroot 'Monogram'

Beetroot 'Detroit 2 Little Ball'

Spinach 'Medania'

Spinach 'Monnopa'

Perpetual spinach

Chard 'Bright Lights'

Chard 'Bright Yellow'

CUCURBITS see pages 85–88

Courgette 'Defender'

Courgette 'El Greco'

Courgette 'Jemmer'

Courgette 'Rondo di Nice'

Courgette 'Orelia'

Marrow 'Long Green Trailing'

Marrow 'Tiger Cross'

Summer squash 'White Patty Pan'

Summer squash 'Sunburst'

Pumpkin 'Atlantic Giant'

Cucumber 'Tokyo Slicer'

Cucumber 'Bush Champion'

Cucumber 'Burpless Tasty Green'

Cucumber 'Marketmore'

'Imai Early Yellow' (AGM) – Flattened globes in a dark straw colour.

**'Senshyi Yellow'* – Straw-coloured bulbs. A good cropper.

Seed to harvest: 20–24 weeks.

SHALLOTS FROM SETS

'Golden Gourmet' (AGM) – Yellow-skinned, pear-shaped shallots. Stores well. HDRA trials show it can be planted from February.

'Hative de Niort' – Pear-shaped bulbs with a dark skin – a favourite for exhibition.

'Pikant' (AGM) – The bulbs are well protected by many layers of reddish-brown skins. HDRA recommends it as bolt resistant.

**'Sante' (AGM)* – A big round shallot with brown skin and pinkish flesh. Stores well and yields prolifically. Mild taste.

'Topper' – Round golden bulbs that store exceptionally well. It was the most reliable in the last GW trial.

Sets to harvest: 18–20 weeks when planted in spring.

PICKLING ONIONS

(SILVERSKIN, COCKTAIL OR BUTTON ONIONS)

Pickling onions are usually grown from seed outside in spring. Plant close together and don't bother to thin them. If you want to keep them white, sow them 5 cm (2 in) deep. They are ready when the leaves die down about two months later.

Recommended varieties

'Jetset' – A mini brown-skinned onion. Matures quickly and is resistant to bolting.

**'Paris Silverskin'* – Pearly white cocktail onions for stews or pickling.

**'Purplette'* – The first purple-skinned variety. Can be harvested early for salads. It turns pink when pickled or cooked.

'Shakespeare' – A reliable, brown-skinned variety.

BUNCHING ONIONS

Allium cepa

The group of plants known as bunching or salad onions are a useful group for salads and stir-fries. They like the same soil and conditions as globe onions.

Spring onions

Spring onions are the most refined types, and some can stand out in winter. Welsh onions are hardier and the newer Japanese bunching onions have some of the advantages of both.

Cooking

The spring onion is generally eaten raw or used as a garnish. They are good chopped in stir-fries (though the Welsh onion is more authentic). The French chop them and add them to soup.

Sowing

Sow seed every two or three weeks for non-stop production, starting in early spring with crop covers and carrying on to midsummer. The hardier types can be sown in late summer for the following spring. Sow thinly in drills about 2.5 cm (1 in) apart. Water in dry periods and pull them up as you need them.

Recommended varieties

'Deep Purple' – A new cultivar with violet, torpedo-shaped bulbs.

'Ramrod' (AGM) – Long, straight white stems. Good for frequent sowing and winter hardy.

**'White Lisbon' (AGM)* – A quick grower for successional crops. Mid-green leaves. Overwinters well.

'Winter White Bunching' (AGM) – A vigorous variety which overwinters well.

Seed to harvest: 12 weeks.

WELSH ONION

(HOLLOW LEEK)

Allium fistulosum

Welsh onions are are a traditional ingredient in Chinese and Japanese cooking. They are nothing to do with Wales – in fact they come from Siberia. It would seem that the name came from the German *walsch* meaning foreign. They are like coarse chives with hollow stems growing to 30–45 cm (12–18 in) tall. They are perfectly happy standing out in the cold and need little attention.

Sowing and planting

Most are herbaceous perennials and die down in winter. The clumps can be dug up, pulled apart, divided and moved around the plot every few years with no bother. Sow them about 23 cm (9 in) apart to give them room to clump up. Modern cultivars are self-blanching.

Spring-sown seeds should be ready by autumn, and those sown in late summer ready the following spring. Keep the area weed free. Pick the leaves whenever you need them.

Recommended varieties

'Welsh Red' – These are hardy plants from Siberia with a strong flavour.

JAPANESE BUNCHING ONION

Allium fistulosum

The Japanese bunching onion is a development of the Welsh onion. It is a more refined and milder tasting version. Depending on variety and how long you leave it to grow, it can be used as a spring onion or cooked like a leek. Although perennial, it is usually grown as an annual in the UK. Seed can be sown from spring and they can be cropped as salad leaves in about six weeks.

They can be pulled out, snipped at any time for a bit of flavour or left in the ground where they will continue to grow for months, finally reaching leek proportions. They can go on through mild winters especially if given some cover. Seedsmen are working on crosses with spring onions. They are less fussy than most onions about soil.

Recommended varieties

**'Ishikura' (AGM)* – A fast grower with long white stems.

'Kyoto Market' – Good for early sowings.

**'Redmate'* – This has an interesting red tinge to the base of the stems.

Seed to harvest: 8 weeks for salads.

GARLIC

Allium sativum

Garlic must be one of the most ancient vegetables known to man and one of the most highly appreciated for its health-giving qualities. As long as it has some sunshine, it is really easy to grow. You need to start off with bulbs from a nursery or seed merchant (not a greengrocer), as they will be certified free of disease and should be suited to a UK climate. After that you can grow from your own stock. Garlic seems to adapt to any given conditions over time.

Garlic types

The type commonly grown in Britain is the non-flowering soft-necked type. The more unusual hard-necked type does flower. It still produces good bulbs, especially if you cut the flowering stem back by half a couple of weeks before flowering. Elephant garlic needs no description. It should be harvested just before the flower opens. Garlic bulbs come in pink, purple and white, and various strengths of flavour.

Cooking

It would be difficult to imagine a world without garlic. It is one of the most essential flavourings across all cultures. Garlic can be used subtly for flavouring sauces, soups, meat and fish dishes, also cheese and bread. It can also be used brazenly in the garlic mayonnaise *aioli*, or the Moroccan *harissa*, made of chilli, garlic, herbs and oil.

In the south of France they have garlic and basil fêtes (the combination with parmesan is the basis of *pistou*) to celebrate the onion harvest. Young cloves of garlic are pickled in India with curry leaves, chilli and mustard seed. In Europe whole cloves are eaten pickled or smoked.

The longer garlic is cooked, the milder it becomes. Slowly roasted garlic cloves become meltingly soft. The French dish of chicken roasted with 40 garlic cloves is very mild. Garlic can be eaten green or dried for later use. It is known to lower cholesterol and to be antibiotic. It contains vitamin C, potassium and iron.

Soil and situation

Plant in a light, sandy soil in a sheltered sunny spot.

Sowing and planting

In the UK, garlic is planted in autumn in warmer areas or in late winter in colder ones. It needs a month or two of cold weather at 0–10°C (32–50°F) to make bulblets. Choose plump, healthy-looking cloves. Split them up and discard any weaklings. Plant with the basal plate (flat end) facing down, 7.5–10 cm (3–4 in) deep, using a dibber. Space 30 cm (12 in) apart.

Cultivation

They will need little attention apart from weeding and watering in dry spells.

Harvesting

Unlike bulb onions, you need to harvest garlic when the leaves go yellow but before they dry out too much. Dig them up, taking care not to bruise them. Dry outside or in an airy shed for a week or so. Don't forget to keep a few healthy ones back for replanting.

Recommended varieties

**'Cristo'* – Large bulbs with up to 15 cloves.

**'Early Wight'* – Adapted to the British climate on the Isle of Wight. An early purple variety.

**'Elephant'* – Giant bulbs with a mild, sweet flavour.

'Thermidrome' – Selected for the UK climate. Plant in November.

'White Pearl' – This has a strong resistance to virus, white rot and eelworm.

Cloves to harvest: variable, 16–36 weeks.

LEEK

Allium porrum

The leek is a five-star allotment plant. While particularly associated with Wales, it is also celebrated in the north-east of England, an area famous for the pot leek competitions that still take place in many a pub and village hall. The name comes from the Anglo-Saxon *leac*. They are robust plants and easy to grow.

Leek types

The earlies are for harvesting in late summer. They are slim, tall and less hardy than the others. Maincrop leeks are for winter eating, while the lates are ready in spring. There are slim or stout types.

Cooking

The leek is a lovely winter vegetable, sweeter and milder than onion. It is the staple ingredient of many classic soups like *Vichyssoise*, *soupe à la bonne femme* and Cockie-leekie; and pies like chicken and leek. Leeks go wonderfully with cheese sauce. Young leeks can be eaten cold in the Greek style with an olive oil dressing.

Soil and situation

Leeks are in the ground for a long time so it's important to get the soil right. It should be fertile, well-drained, on the light side and manured the autumn before. Leeks prefer an alkaline soil, pH7 or above. Lime if necessary.

Sowing and planting

One positive advantage to starting leeks off in a nursery bed or in modules is that you can plant them more deeply to blanch the stems than if you sowed in situ.

Earlies Sow seed in a propagator set to about 12°C (52°F) in late winter. Grow on until the young leeks are about 20 cm (8 in) tall and ready to be hardened off for transplanting. Make holes where they are to grow, about 15 cm (6 in) deep, with a dibber and drop the leeks in. Contrary to common practice, don't trim the roots. Water them in gently.

As they grow, earth them up bit by bit to keep the stems white. Try to avoid getting soil between the leaves. If you grow in situ, put collars (small sections of plastic pipe or tubes of roofing felt or cardboard) around the necks to block out the light as they grow to keep them white. Space them about 15 cm (6 in) apart. If you want big leeks, increase the spacing.

Maincrop Sow in spring in modules in a cold greenhouse, cold frame or nursery bed when the temperature is at least 7°C (46°F). Warm the soil if necessary beforehand with polythene. Cover with cloches in cold weather. Carry on as for earlies but space maincrop leeks more widely.

Lates These follow on from the maincrops. Sow a couple of weeks later.

Cultivation

Keep weeded and watered until they are established. After that they need little attention.

Harvesting

Dig them up as you want them from late summer. If you need the ground for other crops at the end of the season, they will stay fresh for a few weeks if lifted and heeled in elsewhere.

Problems

Can be affected by all onion pests and diseases.

Recommended varieties

EARLIES

'Carlton' F1 (AGM) – Good uniformity. Slow to bolt. Recommended by GW.

*'*Jolant' (AGM)* – This has a long season of cropping for an early type.

**'King Richard' (AGM)* – An abundant cropper, long white shanks and a mild taste. Resistance to bolting.

MAINCROP

'Autumn Mammoth – Cobra' (AGM) – Mid-season to late. Medium-length leeks, bolt resistant.

'Bleu de Solaise' – A French variety with bluey leaves and a long cropping season.

**'King Richard'* – Widely recommended for mini leeks. Plant close together.

'Mammouth Blanche' (AGM) – High yields of big heavy leeks. Often used for showing.

'Swiss Giant – Jolant' (AGM) – December cropping. Good yields with long, solid shafts. Few bolters or 'bulbers'.

**'Toledo' (AGM)* – Early winter to late spring. Uniform and smooth.

'Upton' F1 (AGM) – Has good rust resistance.

Seed to harvest: 16–20 weeks.

THE TREE ONION

(EGYPTION ONION)

Allium cepa Prolifera Group

The tree onion is the joker in the pack. It bears little bulbs on top of a stiff stem. If planted in the spring, the following year it will produce hazlenut-sized onion bulbs. Eventually the stem will bend over and the little bulbs will root themsleves. These can be pickled (or replanted) and the larger onion bulbs below ground can be eaten as salad onions. It is really only grown as a curiosity as yields are low and the bulbs hard to find.

BEETROOT, SPINACH AND OTHER GREENS

The ancient ancestor of the beetroot family, *Beta vulgaris*, is a coastal plant found in Europe, North Africa, the Middle East and Asia. The spinach-like leaves have been enjoyed as a vegetable since prehistory. Wild beet was recorded in the Hanging Gardens of Babylon and was offered to Apollo at the temple in Delphi. From it came sugar beet for sugar, mangold for cattle feed as well as beetroot, true spinach, perpetual and New Zealand spinach, orache, chard and leaf beet.

True spinach is considered to be the most refined of the group. As luck would have it, it is the only one that is tricky to grow, but it makes an excellent cut-and-come-again crop. New Zealand and perpetual spinach and the chards are easy and crop over a long period. Some are very eye-catching.

All need light, sandy, fertile soil, preferably manured for the previous crop. They prefer neutral to alkaline soil, pH7–7.5. Problems are generally few as all grow too fast for anything to get a hold.

Problems

Downy mildew, leaf spot fungi, shot hole, magnesium deficiency, cutworms, slugs, snails and birds could affect the plants. Spinach beet and chard can get leaf miner or beet leaf spot.

BEETROOT

Beta vulgaris subsp. *vulgaris*

The Romans cultivated wild beet, selecting the sports with the biggest roots to grow on for medicine. However, it wasn't until the 16th century that beetroot was really developed as a vegetable. It would seem most of the breeding took place in Italy, as it was known as Roman beet at first. The Germans latched on, developing and promoting it, and it became a staple food in Central and Eastern Europe. Borscht, the classic beetroot soup widely enjoyed in Russia, Poland and the Ukraine, dates from this time.

Closer to home in 1597, John Gerard seems to find beetroot something of a novelty. He says, it "hath leaves very great, and red of colour, as is all the rest of the plant, as well as root, as stalke, and floures full of perfect purple juyce tending to redness." He recommends the leaves for winter salads as being both "pleasant to the taste, but also delightful to the eye". He seems rather mystified on how to deal with the root however. He suggests that "the cunning cooke" would come up with some good ideas.

Beetroot types

Beetroot these days comes round, globe shaped, tapered, flat and oval. There are varieties that are white, gold, purple, red, even one striped like raspberry ripple. There are fast growers for spring, and slow ones that store better for the winter months.

Cooking

In the UK, young beetroot is usually eaten cold as a salad ingredient, boiled or raw and grated. To keep in flavour, beetroot can be baked in its skin in the same way as a potato. In Scandinavia, *sildesalat*, or herring and beetroot salad, is a traditional dish. In Holland, it is puréed with apples, onions and nutmeg. The classic Russian *borscht* is served either hot or cold topped with sour cream. Beetroot can be curried, made into chutneys or pickled. It is rich in potassium and calcium.

Soil and situation

Soil as above. Long-rooted types need a good depth of top soil. Though not essential, many recommend a dressing of agricultural salt as beet is a seaside plant. A sunny and open situation is required.

Sowing and planting

Seed usually comes in clusters of several seeds. As this involves thinning out, monogerm, or single seed types, are a popular choice. They benefit from being soaked to soften the coating for half an hour before sowing. It is best to use bolt-resistant cultivars for spring sowings. They can be sown outside when the soil has reached a minimum of 7°C (46°F). Sow about 2 cm (3/4 in) deep, 2.5 cm (1 in) apart, in rows 20 cm (8 in) apart. Thin to a final spacing of 10 cm (4 in). Sow a few seeds every fortnight for continuous crops. Switch to maincrop types in June.

Cultivation

Make sure they don't get too wet or too dry. Too much water will make them put out more leaf than root. On the other hand, if allowed to dry out, the roots will toughen. If suddenly soaked when dry, the roots may fork. Mulch to keep in moisture. Weed carefully as any injury to the roots will make them bleed.

Harvesting

Twist off the leaves about 5 cm (2 in) above the root to avoid bleeding. Loosen the soil with a fork and pull up the roots carefully to avoid damage. For succulent beetroot, harvest earlies when they are the size of golf balls, and maincrops when the size of a tennis ball. Lift the remaining late crops in autumn and store in moist sand in a cool shed. In areas where the temperature is slightly milder, the late ones can be left in the ground until wanted under a good covering of straw or bracken.

Problems

Usually trouble free, but may bolt.

Recommended varieties

★'Boltardy' (AGM) – A good bolt-resistant variety excellent for early sowings. Round red type. Popular choice.

'Bonel' (AGM) – Bolt resistant, early cropper. High yields of round, red, tasty roots.

★'Bulls Blood' – An old variety with dramatic red foliage.

★'Burpees Golden' – A yellow variety with orange skin. Doesn't bleed when cut. Tasty.

'Cheltenham Green Top' (AGM) – An old variety with a tapered root.

'Cheltenham Mono' – Monogerm and resistant to bolting.

★'Chioggia' – Roots are red with white stripes. Small with exceptionally sweet flesh.

'Moneta', *'Monopoly'* and *★'Monogram'* – All monogerm types, which saves thinning.

'Regala' (AGM) – Good resistance to bolting. Stores well. Recommended.

MINI VARIETIES

★'Detroit 2 Little Ball', *'Pronto'* and *'Monaco'* are all baby beets.

Seed to harvest: 8–13 weeks.

SPINACH

Spinacia oleracea

True spinach is thought to have come from Persia, reaching the UK in the 16th century. It became widely grown in European monastery gardens. Spinach is a cool climate crop and likes damp conditions. Its reputation for being temperamental is due its tendency to bolt, though modern breeding has curbed this. You need quite a few plants to get any quantity as spinach shrinks dramatically when cooked.

Spinach types

Traditionally the summer varieties are smooth-seeded and are considered superior to the tougher prickly-seeded types. Modern breeding has produced seeds that can be used for both winter and summer. They have reduced the bitterness, the tendency to run to seed at the slightest hitch, and have introduced red-leaved varieties. The leaves are either round or arrow-shaped, smooth or crinkled.

Cooking

Spinach is an international vegetable. It is classically cooked lightly stewed in its own juice and partnered with good oils, cheeses, butter and a little nutmeg to bring out the flavour. It goes well with eggs. Classic dishes include *oeufs Florentine*, stuffed pancakes, soufflés and roulades. In Persia, they make a Spanish-style omelette with spinach and potato known as *kukuye esfanaj*. In India potato and spinach is combined for *saag gosht*. Spinach is useful for filling pasties, pies, gnocchi and pasta. In France they make a sweet spinach tart, *tarte d'epinards au sucre*. Spinach makes silky soups and the young leaves are good in salads. Spinach is rich in vitamins A, C and B2, plus iron.

Soil and situation

Spinach likes moisture-retentive soil, neutral to alkaline, rich in organic matter. Choose a sunny position in spring, part shade in the heat of summer.

Sowing and planting

For the first sowings, choose bolt-resistant varieties. If you want early crops you can raise them in modules or grow through plastic on pre-warmed soil, but it is easier to sow them in situ in late spring. Soak the seed overnight. Sow thinly about 2 cm (3/4 in) deep. Thin alternate plants, aiming to end up with 15 cm (6 in) between them. The thinnings can be used in salads. Rows should be about 20 cm (8 in) apart. For a succession, sow every two weeks.

Winter varieties are sown in August or September for the following spring. Sow a few batches at two week intervals as the timing is something of a guessing game. You want them to be large enough to survive the winter but not to have bolted before the cold weather arrives.

For cut-and-come-again crops, sow thinly every couple of weeks from spring to summer, 10 cm (4 in) apart.

Cultivation

Don't let spinach dry out, ever. Give high-nitrogen liquid feed every two weeks. The winter ones will need protection.

Harvesting

Pick a few outside leaves constantly before they toughen, going from plant to plant. Never take more than 50 per cent so that the plants can recover and put out new leaves.

Problems

Bolting, downy mildew, birds.

Recommended varieties

'Atlanta' (AGM) – Frost resistant with thick, dark green leaves. High yields.

★'Medania' (AGM) – Vigorous with big leaves. Resistant to bolting.

★'Monnopa' (AGM) – Autumn or summer, vigorous yet slow to bolt. Thick leaves.

'Palco' (AGM) – Mildew resistant, slow to bolt. Late.

'Sigmaleaf' (AGM) – Proven to be hardy throughout the British Isles. Good crops. Tasty.

'Spokane' F1 (AGM) – Mildew resistant, slow to bolt. Stands well.

Seed to harvest: 2 weeks for cut-and-come-again; 8–10 weeks to maturity.

NEW ZEALAND SPINACH

Tetragonia tetragonioides

This is the spinach that Captain Cook ordered for his crew to prevent scurvy. It was introduced to Kew Gardens by Joseph Banks, the famous plant hunter in 1772, after his voyage to New Zealand on the Endeavour. Unlike true spinach, it is easy going and takes the heat of summer without protest. As it is not hardy, it is grown as an annual in the UK. It's a low and sprawling plant that needs a bit of space – perhaps this is the reason that it is not widely grown in gardens. It makes a good allotment plant, though, with the added bonus of providing weed cover.

New Zealand spinach types

No cultivars.

Cooking

The leaves, if picked young, are creamy rather like sorrel. They can be used in the same way as spinach. The midribs and seeds are not eaten.

Soil and situation

New Zealand spinach grows wild on the sandy beaches of Australasia and the Pacific Islands. Its needs, therefore, are damp sandy soil and sun. It tolerates poor conditions.

Sowing and planting

Wait until the frosts are truly over. Soak seed overnight and station sow two or three seeds in situ about 2 cm (3/4 in) deep and 45 cm (18 in) apart and between rows. Thin to the strongest when big enough to handle. Alternatively you can grow them under cover in mid-spring and harden them off carefully before transplanting in May or June.

Cultivation

Pinch out the growing tips to encourage side shoots. Weed until the plants are established. Although they tolerate drought, keeping them watered will improve the crop. They self-seed easily.

Harvesting

Harvest from late June to September, regularly picking the outer leaves before they toughen. The more you pick the more will come. Eat fresh.

Problems

Usually trouble free.

Seed to harvest: 6–7 weeks.

SPINACH BEET

(LEAF BEET OR *PERPETUAL SPINACH)

Beta vulgaris Cicla Group

Spinach beet may be more peppery and less subtle than spinach but it has the advantage of being reliable and easy to grow. It doesn't bolt easily, it is more winter hardy than true spinach and resilient enough to withstand seaside conditions in the UK. This is the one for a windy allotment. It is not grown commercially as it doesn't have a good shelf life. Excellent for cut-and-come-again crops, it also makes for good winter greens outside under cover. Seeds sown in spring will keep going right through summer and, with luck, through winter.

Spinach beet types

No cultivars.

Cooking

As for spinach.

Soil and situation

As for spinach.

Sowing and planting

The seeds come in clusters. Start off in late spring. Soak seeds and sow thinly 2.5 cm (1 in) deep and about 20 cm (8 in) apart. Thin to the strongest. Sow in March or April outside for summer and autumn croppings. A second sowing in mid to late summer will give you fresh spinachy leaves through winter and into the hungry gap. Provide covers to get the best crops. For cut-and-come-again, sow more closely, 10 cm (4 in) apart.

Cultivation

Keep watered in summer. If the plants start to flag, give them a high-nitrogen feed. If they don't pick up after this, or if they coarsen, cut them down to the ground and they will put out fresh growth.

Harvesting

Continually pick leaves from the outside when young and succulent, never taking more than 50 per cent to encourage fresh production.

Problems

Generally trouble free. Leaf spot on older leaves.

CHARD

(SWISS CHARD, RUBY CHARD, SEAKALE BEET, SILVER CHARD, SILVER BEET)

Beta vulgaris Cicla group

The chards are the glamorous side of the beet family. The red-stemmed types are almost luminous when backlit by the sun. They produce over a long period and are trouble-free given the right nurture.

Chard types

Gardeners are spoilt for choice. They can be smooth or crinkly; with white, crimson, yellow or purplish stems and green, reddish or rainbow-coloured leaves.

Cooking

The white-stemmed chard with large crumpled green leaves is considered to be the tastiest type. The chards are greatly enjoyed in Provençal cookery for rice dishes, soups and for special tarts with nuts and fruits, lemon and cheese. Sadly, the brightly coloured varieties lose their colour when cooked. The midribs are usually cooked separately and eaten with a vinaigrette dressing.

Cultivation

They are grown in the same way as spinach beet.

Recommended varieties

**'Bright Lights' (AGM)* – The rainbow chard.

**'Bright Yellow' (AGM)* – Yellow stems with a puckered green leaf.

'Charlotte' (AGM) – Bright red.

'Fordhook Giant' (AGM) – White ribs and green leaves.

'Lucullus' (AGM) – Good flavour.

'Rhubarb Chard' (AGM) – Scarlet ribs. Purplish, puckered leaves.

Seed to harvest: 8–13 weeks.

QUINOA

Chenopodium quinoa

Quinoa is a plant with presence, growing 1–1.5 m (3–5 ft) high. Primarily grown for its grain, it produces spinach-like leaves as well. The plume-like seedheads are orange, red, yellow, black or white. Perhaps it is not a serious proposition for the allotment holder, though it would certainly make a talking point and an ideal ornamental summer screen.

A fast-growing annual, it is occasionally grown in the UK as a cover crop and grain crop for game. Quinoa has formed a vital part of the Andean diet for 5,000 years. The name comes from *quinua* which means mother grain. It has been ground into flour and made into alcohol since the Inca civilization. It grows wild in the Andes at 4,000 metres.

Quinoa types

There are early and maincrop varieties in varying heights and colours.

Cooking

The seeds come in a bitter seed coat that needs to be removed. Once shucked, they are thoroughly washed until the water is clear, then boiled like rice until transparent. They have a rather delicious nutty taste and texture. They can be eaten in the same way as couscous or toasted.

The young leaves are green,

turning to red or purple, and are prepared in the same ways as spinach. Quinoa is not called mother grain without reason. It is a super food with more vitamin A and E than spinach. It is a complete protein and it contains manganese, magnesium, iron, copper and phosphorous.

Soil and situation

Soil as above, though it can cope with poor conditions. Avoid a windy site.

Sowing and planting

Quinoa needs a period of cold before it will germinate, so put the seeds in the fridge for a few days. It won't germinate in warm soil. Sow in situ in mid-spring when the soil is 15°C (59°F), about 2 cm (3/4 in) deep and 40 cm (16 in) apart. If you want the plants to grow as big as possible, space them more widely. They look good in a block formation. Cover the seed lightly.

Cultivation

Weed around the young plants carefully until they establish. Quinoas don't need much water or attention, being a tough field crop.

Harvesting

The seeds are ready for harvesting when they are all the same colour. The stems are cut down and dried hanging upside down. In the Andes, the farmers stack them up to make an arch for drying. When they are ready, they loosen the seed heads by beating them with sticks before winnowing the seed.

Problems

The most likely are wrong weather conditions – a wet or scorching hot summer. Pests dislike the grain due to its bitter seed coats.

ORACHE

(MOUNTAIN SPINACH)

Atriplex hortensis

Once a popular crop in Central Europe, orache has become something of a novelty these days. Orache has a lot going for it – it is really easy to grow and very ornamental. The wild form is the common weed fat hen (*Atriplex patula*). The cultivated form grows extremely tall, up to 2 m (6 ft 8 in), at speed.

Orache types

The small triangular leaves come in claret, gold or green, according to variety.

Cooking

Cook the leaves like spinach or eat the young ones raw in salads. The red-leaved variety *Atriplex hortensis* var. *rubra* looks particularly attractive.

Soil and situation

Though not fussy, a good rich, free-draining, sandy soil will produce the best crops. Coloured varieties may scorch in the sun so find them a spot in semi-shade.

Sowing and planting

Orache germinates easily. Unless you want a very early crop, don't bother with sowing under cover. Sow outside in mid-spring about 25 cm (10 in) apart. It usually goes to seed by midsummer, so for longer supply, make a second sowing a few weeks later.

Cultivation

Keep well watered. To prolong the harvest, remove the flowers as soon as you see them unless you want to keep the seed. Orache seeds itself as prolifically as a weed.

Harvesting

Pick the leaves when young and don't let it seed.

Problems

Usually problem free.

Recommended varieties

'Green Spikes' – Produces abundant pale green leaves.

'Opera Red' – A wonderful, old, deep red variety which retains its colour when cooked. Lovely in salads.

Seed to harvest: 6–8 weeks.

AMARANTHUS

(CHINESE, AFRICAN OR INDIAN SPINACH, BAYAM, CALALOO)

Amaranthus gangeticus

Amaranthus is vigorous and easy to grow, providing you keep in mind that it is not hardy. The history of the grain amaranthus (*A. cruentus*) goes back to the pre-Columbian Aztecs. It is grown in China, India, Africa, Europe and the Americas as a cereal crop – a striking one with magnificent feathery plumes like corn tassels in red or magenta. In the UK, it is usually harvested young for its leaves.

Amaranthus types

There are large- and small-leaved types. Colours vary from light to dark green, red and also variegated.

Cooking

The leaves are a classic accompaniment to salt fish in West Indian cuisine. The youngest leaves are good in salads.

Soil and situation

As above. Although they are not too fussy, good fertility will reward you with the best crops. Choose a warm, sheltered site.

Sowing and planting

The lowest soil temperature for germination is 10°C (50°F). A higher temperature is better, so wait until the soil is warm in late May and sow straight outside. As the seed is very fine, mix it with a little sand or sieved soil. Sow about 2 cm (3/4 in) deep and cover well to cut out the light. Provide extra warmth with crop covers. When the seeds pop up, thin to a spacing of 10 cm (4 in).

Cultivation

Keep watered and feed every couple of weeks with liquid seaweed. Pinch out the growing stem when about 20 cm (8 in) high to encourage the plants to bush out. Take great care not to let the plants flower – they seed prolifically and can become a nuisance.

Harvesting

Harvest when young and before the leaves coarsen. Either pick off the outer leaves or pull up the entire plant with its roots when about 20–25cm (8–10 in) high. The leaves wilt quickly so harvest them for eating straightaway.

Recommended varieties

Not widely available.

Seed to harvest: 10–12 weeks.

CHRYSANTHEMUM GREENS

(CHOP SUEY GREENS, GARLAND CHRYSANTHEMUM, JAPANESE GREENS, SHUNGIKU)

Chrysanthemum coronarium

Chrysanthemum greens are a form of the annual garden chrysanthemum. They are widely used in Japanese cuisine. The leaves are picked when about 10 cm (4 in) tall before they become bitter and strong tasting. If you forget to harvest them in time, you can still enjoy the flowers.

Chrysanthemum greens types

There are varying leaf sizes. No named varieties in the West.

Cooking

The young leaves are added to salads or lightly cooked in soups, stews or used for tempura.

Soil and situation

They are not fussy, but the best results are achieved on rich, moisture-retentive soil. Choose an open situation with sunshine, or light shade in summer.

Sowing and planting

You can sow seeds throughout the growing season, though late summer is the best time for cropping up until the frosts. Sow the fine seed as sparsely as possible on a carefully raked and moistened surface. Barely cover. Thin to 15 cm (6 in) apart for individual plants. They work well as a cut-and-come-again crop as they regenerate easily. You can sow them in modules for transplanting if you prefer, or grow them through winter under cover.

Cultivation
Keep weed free. Pinch out the tips to get bushy plants.

Harvesting
Pick the leaves when the plants are about 10 cm (4 in) tall or pull up the entire plant.

Problems
None.

Seed to harvest: 8 weeks through the growing season.

GOOD KING HENRY
(ALL-GOO, LINCOLNSHIRE SPINACH)
Chenopodium bonus-henricus
Good King Henry is an old fashioned perennial from the beetroot family. You won't find it in the shops as it wilts quickly after picking but it is a great plant for the allotment. Introduced to Britain by the Romans to feed the legions, it is undemanding. It grows to 90 cm (3 ft) tall.

Cooking
The arrow-shaped leaves are eaten in the same way as spinach and the flowering shoots can be enjoyed like asparagus.

Soil and situation
Though Good King Henry will survive anywhere, to get the best from the crop find it a well-drained and fertile patch that is partially shaded in the summer months.

Sowing and planting
Start off from seed. Sow in situ in spring 1 cm (1/2 in) deep. Thin to 20 cm (8 in) apart when big enough to handle. They can be grown under cover without heat for an early crop if so desired. Once you have established plants, divide them in spring (see page 155).

Cultivation
Water in the heat of summer.

Harvesting
Don't crop until the second year. The spring after sowing, snip off the flowering spikes and later in the year pick the outer leaves as you want them. It is advisable to split the plants every third year.

Problems
Unlikely to arise.

Seed to harvest: two years, then harvest every year.

CUCURBITS

Courgettes, marrows, summer squash, pumpkin, winter squash, cucumber and gherkins belong to the same family of half-hardies from warm countries. The earliest records of pumpkin and squash have been traced back 7,000 years to Mexico.

They are vigorous, bushy, trailing annuals. They grow so fast that they can easily be grown outside to fruit in a British summer, especially if started off indoors using the new hardier varieties. They are fun to grow and, bar cucumbers, a great first crop for children to try.

Cucurbits prefer soil on the acid side, pH6–6.5. They have male and female flowers on the same plant. You will need to remove any covers when they are in flower for the insects to pollinate them. They need sunshine and shelter, pre-warmed soil, cloche cover in cool parts of the country (with ventilation in summer to avoid scorching) and a good supply of nutrients and water. Put in plenty of well-rotted compost, or you can even grow them on the manure or compost heap. They are so thirsty that they benefit from having a piece of open-ended pipe or sawn-off plastic bottle inserted into the ground to get water down to the roots.

COURGETTE, MARROW AND SUMMER SQUASH

(ZUCCHINI)

Cucurbita pepo

Courgette, marrow and summer squash have long been an important food crop in North and South America. The courgette is just a young marrow, so you can get two crops in one if you let a few grow on to maturity. Summer squashes come in wonderfully ornamental shapes. All should be eaten soon after they ripen as they don't keep for long.

Courgette, marrow and summer squash types

Courgettes and marrows have round or long fruits in green or yellow. The bush types produce the best courgettes for intensive cropping.

Summer squashes fall into four main groups – scallop (or custard) squash, crookneck, straightneck and vegetable spaghetti.

Cooking

Courgettes If you grow them yourself, you can pick and eat them raw when really tiny. They are good with dips. If you plan to fry courgettes, salting them in the same way as aubergines will make them less absorbent of oil. In Greece, they serve courgette fritters with yoghurt. They can be grilled or griddled on an open fire in slices. They are often cooked with tomatoes, handy as they usually mature at the same time. The flowers of courgettes are a delicacy deep-fried in batter.

Marrows As the courgettes coarsen into marrows, they can be blanched, deseeded and baked with stuffings.

Summer squashes These can be treated in the same way as courgettes. Vegetable spaghetti can be baked either whole or split in half. If the centre is pulled out and separated with a fork it will split into pasta-like strands. It is good eaten with pasta sauces.

Soil and situation

Soil as above. Choose a warm spot.

Sowing and planting

Soak the seeds overnight. Sow them about 2.5 cm (1 in) deep on their sides in individual pots indoors in late April or early May at 18°C (64°F). Harden off carefully after all danger of frost has passed and plant outside under a cold frame. Alternatively, sow outside in late May on pre-warmed soil under cloches.

Cultivation

The climbing types will need strong supports. You can grow them up canes or posts, pinching out the top when they reach the desired height. Feed at least once a fortnight with liquid feed and keep well watered. Growing them through black polythene will give them extra warmth and keep the fruits of the sprawling types clean.

Harvesting

Harvest the fruits by cutting them off with a sharp knife when young and succulent to encourage more to come. Catch the marrows before they coarsen. The courgettes and summer squashes will be ready from July onwards.

Problems

They can suffer from powdery mildew, cucumber mosaic virus, slugs and snails. In the cold frame, watch for red spider mite and whitefly.

Recommended varieties

COURGETTES

'Bambino' F1 (AGM) – Small, tender fruits. Bush type. Early and prolific. A popular choice.

**'Defender' (AGM)* – Dark green, lightly flecked fruits. Resistant to cucumber mosaic virus.

'Early Gem' F1 (AGM) – Dark green slim fruits. Prolific.

**'El Greco' F1 (AGM)* – Bush type with mid-green fruits.

**'Jemmer' F1 (AGM)* – A prolific yellow variety with compact fruits.

**'Rondo di Nice'* – An Italian round courgette.

MARROWS

'Badger Cross' F1 (AGM) – Also resistant to cucumber mosaic virus.

'Clarita' F1 (AGM) – Pear-shaped fruits and high yields.

**'Long Green Trailing'* – A prolific, traditional striped marrow.

**'Tiger Cross F1 (AGM)* – A late bush type resistant to cucumber mosaic virus.

SUMMER SQUASH

'Patty Pan' and **'White Patty Pan'* – Both have a pretty scalloped edge. Very ornamental.

**'Sunburst' F1* – Yellow, scalloped, flying saucer-shaped fruit.

'Vegetable Spaghetti' – Pale yellow. When cooked the flesh inside looks like spaghetti.

Seed to harvest: courgettes and summer squash 6 weeks; marrows 7–8 weeks.

PUMPKIN AND WINTER SQUASH

Cucurbita maxima

Pumpkins and winter squash develop a hard skin and are usually grown for winter storage. They can also be eaten when immature like courgettes.

Pumpkin and winter squash types

Pumpkins are possibly the largest vegetables in existence. Some varieties can easily reach 100 kg (224 lb) in weight. Autumn pumpkins are round or lozenge-shaped and usually ribbed.

Winter squashes have real character – a pantomime poetry about them. Types include the turban, the warted hubbard, buttercup and banana. They are mostly large, sprawling plants though there are bush varieties. They usually have white or pale yellow flesh.

Cooking

When cooked, the flesh turns into a creamy purée. In America, pumpkin pie is the classic dish for Thanksgiving.

In the UK, pumpkins are traditionally made into soup for Halloween. Cut into chunks, they are roasted in the same way as potatoes. If sliced they can be frittered. The early American settlers baked them whole, seeds removed, over a camp fire wrapped in cabbage leaves. In Germany they combine pumpkin with apple in the same way as they do with red cabbage. The seeds can be salted and baked.

Soil and situation

Rich cucurbit soil.

Sowing and planting

As they take between three and five months to mature, sow around the end of April indoors in individual pots. Plant out in the same way as courgettes and marrows, allowing extra space between plants around 1.2 m (4 ft). In mild areas, they can be sown in situ after the frosts.

Cultivation

They make a strong root system so, once established, they are not as thirsty as courgettes and marrows. Put straw or a board under the fruits to keep them clean.

Harvesting

Let them ripen in the ground until they sound hollow when tapped. Dry them further in the sun, underside up, to cure the skins before storing.

Recommended varieties

PUMPKINS

**'Atlantic Giant'* – Truly gigantic. The world-wide record-breaking, exhibition pumpkin.

'Triple Treat' – Perfect for Halloween. Not too big, it is round and orange with tasty flesh. The seeds are great for toasting.

WINTER SQUASH

'Butterball' – Bred to mature in three months, so a good choice for the UK. Sweet-tasting orange flesh.

'Crown Prince' – A more modest character with grey skin but delicious, nutty orange flesh.

'Crown of Thorns' – Spherical fruits with spikes.

'Turk's Turban' – The name describes the shape. Possibly the most exotic looking of all the winter squashes with orange, cream and green splotches and stripes. Good for soup.

Seed to harvest: 3–5 months.

CUCUMBER AND GHERKIN

Cucumis sativus

The cucumber is another plant of antiquity. It was cultivated from the form that grows wild in northern India. The two main sorts are greenhouse types and the ridged ones for outdoor growing. The greenhouse types are longer, more elegant and have smoother skins than their country cousins. They are also trickier to grow and more susceptible to pests and disease. However, modern breeding of Japanese and Burpless cucumbers has made it possible to grow swish hothouse types out in the open. Gherkins are small cucumbers grown for pickling, though they can be eaten fresh. The cucumber is a tender plant with a very high water content – over 95 per cent.

Cucumber and gherkin types

For growing outside, there are both climbing and bush types. Oddities include round cucumbers. There are white- and yellow-skinned ones as well as various greens.

Cooking

In the west, cucumber is a very traditional salad vegetable or garnish. It is an English classic served in dainty sandwiches for tea in summer while cricket is played on the village green. In Indian cookery it is made into *raita* – grated and mixed into yogurt with chopped mint, it is a cooling accompaniment to curries. It is also mixed with chopped mint and iced sugar water as a cold drink. Cucumber is good in chilled soups. In Spain it is made into gazpacho with tomatoes and garlic, and in Turkey into *cacik* – a cucumber soup with yogurt. In Russia they make a cold cucumber soup with sorrel and chopped boiled eggs. The French make hot cucumber soup with cream and dill.

Soil and situation

Cucumbers like good and rich soil with plenty of well-rotted manure or compost incorporated. A couple of weeks before the plants are ready to go out, dig a trench, half fill it with well-rotted manure or compost and pile back the soil on top to make a ridge (hence ridge cucumbers). Choose a sheltered and sunny position.

Sowing and planting

Sow the outdoor types indoors at 20°C (68°F) in biodegradable modules in late spring, about 2.5 cm (1 in) deep. Time it to a month before the last frosts. Sow two seeds per module and thin to the strongest. Keep the seedlings warm, a minimum 16°C (61°F) at night. Don't be too generous with watering at this stage as they are prone to damping off. Harden them off carefully when they have about three leaves before planting out under cloches. Plant a little less deeply than before to avoid neck rot.

Another option is to sow straight outside in June (or when the soil temperature is at least 20°C/68°F).

Climbing types can be grown on the ground with straw under the fruits, or trained up a wigwam, trellis, wires or netting. This is more practical as it has the double effect of keeping the fruits clean and protecting them (to some extent) from slugs.

Cultivation

Stop the plants by nipping out the growing tip when it reaches the top of the support. Keep well watered, and mulched. When the plants are in flower, remove any covers so the insects can pollinate them. Give liquid feeds, particularly when the fruits form.

Harvesting

Cut the fruits off with a sharp knife when ripe but before they go yellow.

Problems

Problems that can arise are cucumber mosaic virus, slugs, aphids, red spider mite in hot weather, powdery mildew and neck rot.

Recommended varieties

OUTDOOR CUCUMBERS

**'Bush Champion' F1 (AGM)* – A compact variety with high yields. Resistant to cucumber mosaic virus.

**'Burpless Tasty Green' F1* – Mildew resistant. Crisp and well-flavoured fruits. Good croppers. Possibly the best of the new outdoor varieties for glasshouse-quality fruits.

'Crystal Lemon' – An interesting cucumber that looks not unlike a lemon and has a tangy taste. Can be used for pickling as well as eating.

**'Marketmore' (AGM)* – Ridge type, high yielding. Resistant to cucumber mosaic virus, also powdery and downy mildew. Good for cooler climates.

**'Tokyo Slicer' F1 (AGM)* – Japanese variety with slender fruits and smooth skins. Perfect for cucumber sandwiches. Good in our climate.

'Zeina' F1 (AGM) –Vigorous with short tasty fruits.

GHERKINS

'Fortos' (AGM) – A variety with uniform fruits.

'Gherkin' – Fast growing variety with masses of small prickly fruits.

'Vert Petit de Paris' – Prolific, tasty French number.

Seed to harvest: 12 weeks.

FRUITING VEGETABLES see pages 97–103

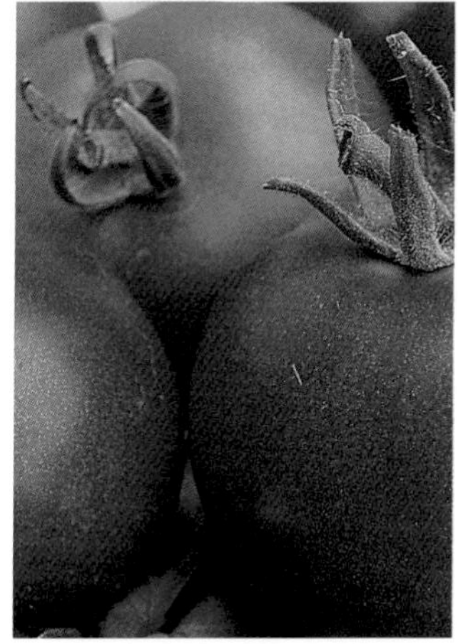

Tomato 'Alicante'

Tomato 'Golden Sunrise'

Tomato 'Shirley'

Tomato 'Tigerella'

Tomato 'Tornado'

Tomato 'Gardener's Delight'

Tomato 'Sweet Million'

Aubergine 'Black Beauty'

Aubergine 'Violetta di Firenze'

Sweet pepper 'Canape'

Sweet pepper 'Gypsy'

Chilli pepper 'Apache'

Chilli pepper 'Jalapeno'

Chilli pepper 'Ring of Fire'

Chilli pepper 'Tabasco Habanero'

Sweet pepper 'Bell Boy'

Sweetcorn 'Dickson'

Sweetcorn 'Sweet Nugget'

Sweetcorn 'Honey Bantam'

Okra 'Pure Luck'

STEM VEGETABLES see pages 103–107

Celery 'Celebrity'

Celery 'Pink Ice'

Celery 'Victoria'

Celeriac 'Monarch'

Florence fennel 'Zefo Fino'

Florence fennel 'Rudy'

SALAD LEAVES see pages 107–113

Lettuce 'Lilian'

Lettuce 'Webb's Wonderful'

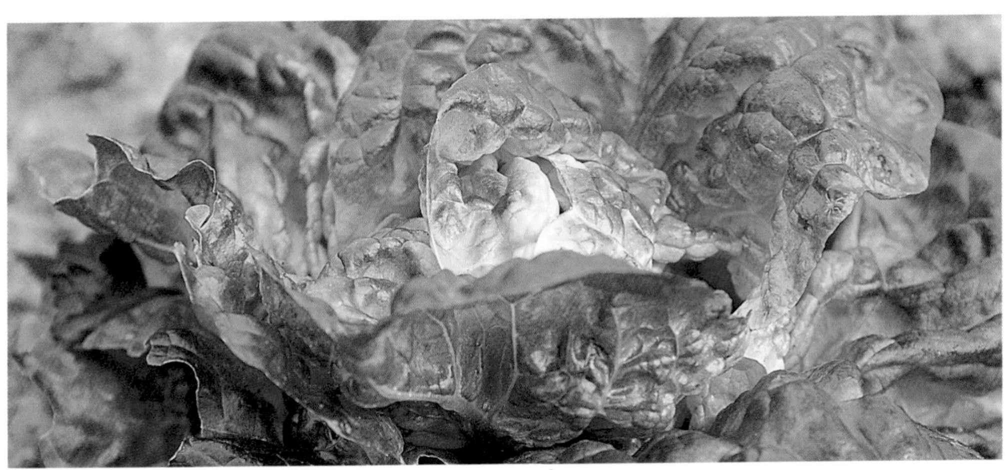

Lettuce 'Merveille de Quatre Saisons'

Lettuce 'Cos'

Lettuce 'Paris White'

Lettuce 'Lollo Bionda'

Lettuce 'Lollo Rossa'

Lettuce 'Oak Leaf'

Lettuce 'Red Salad Bowl'

Lettuce 'Salad Bowl'

Lettuce 'Blush'

Lettuce 'Frillice'

Lettuce 'Little Gem'

Endive 'En Cornet de Bordeaux'

Chicory 'Witloof'

Chicory 'Sugarloaf'

Chicory 'Radicchio'

Chicory 'Palla Rossa'

Chicory 'Rossa di Treviso'

PERENNIALS see pages 114–117

Globe artichoke 'Green Globe'

Globe artichoke 'Purple Globe'

Globe artichoke 'Romanesco'

Globe artichoke 'Violetta'

Cardoon 'Gigante di Romagna'

Seakale 'Lily White'

Wild sea kale

Asparagus 'Jersey Giant'

FRUITING VEGETABLES

Tomatoes, aubergines, peppers, chillis, sweetcorn and okra come from hot countries and need long summers to ripen. Some cold resistance in contemporary F1 seed has made it much easier to grow them outside in Britain in warmer parts of the country. If you grow from seed you will need to provide tender loving care for many weeks, with a minimum temperature of 16°C (61°F) for germination. This can be done in a heated propagator followed by a sunny windowsill. Make sure the temperature doesn't drop below this at night. You can save yourself time and effort by buying young plants from a reputable nursery or garden centre, though you won't have the same choice of varieties.

Choose the sunniest, most sheltered spot and warm the soil with plastic sheeting for a week before planting out. Either grow through slits in the sheets or mulch well to prevent evaporation and keep down weeds. Fruiting vegetables are insatiable and need fertile soil with plenty of organic material incorporated, a general fertilizer applied before planting and top-ups of liquid feed. They also need good drainage. When in flower and fruit, water copiously.

TOMATO

Lycopersicon esculentum

The wild tomato grows in Peru, Ecuador and Chile and was probably first cultivated in Mexico. It is believed that Cortez introduced it to Europe in the 16th century. It was regarded at the time with deep suspicion as, along with the potato, the tomato is another member of the deadly nightshade family, *Solanaceae*.

The Italians seem to have appreciated it first, calling it *pomi d'oro* (golden apple) while the French dubbed it *pomme d'amour* (love apple), believing it to have aphrodisiac qualities. The name tomato comes from the South American name, *tomatas*.

Given the right conditions, tomatoes are well worth growing as, freshly picked from the vine, they are really delicious. Tomatoes are treated as half-hardy annuals in the UK.

Tomato types

The two main divisions lie between vine (indeterminate) and bush (determinate) varieties. The vine tomatoes are greenhouse types. The bush types are a better bet for the allotment as they grow faster and will fit under cloches or crop covers. There is huge diversity in size, from the giant beef tomatoes down to tiny cherry ones. There are wonderful heritage varieties, and a fun range coming in green, yellow, purple, striped and in an assortment of shapes.

Cooking

The tomato is invaluable in the kitchen. It makes a classic marriage with basil in salads, also delicious with finely chopped red onion and parsley. The flavour of really good tomatoes is brought out in the raw chilled Spanish or Arabic classic soup, gazpacho. If you have a glut, follow the French by peeling, chopping and freezing them to make *tomates concassées* for an endless variety of sauces and soups. Green tomatoes can be made into chutney.

Soil and situation

Soil as above, pH5.5–7. Choose a very sunny and sheltered position.

Sowing and planting

Raise the seeds in a greenhouse or propagator, or on a

windowsill, in early April six to eight weeks before the last frost. The minimum temperature for germination is 16°C (61°F). Sow seed 2 cm (3/4 in) deep in seed trays. Pot on when there are three true leaves. Harden off about six weeks later in May when the flowers are just forming and plant out in June when the weather is right. Planting distances will depend on the variety.

Cultivation

Vine tomatoes produce a main stem. It needs to be tied onto a sturdy central support as the plant can become top heavy. As they grow, the side shoots are pinched out and, towards the end of summer, the leader is stopped (or nipped out) to make the plant concentrate less on growing taller and more on producing fruits.

Bush varieties don't need training. Growing in the open air makes them less prone to disease and produces tastier crops. Start them off under cloches until they are well established. Feed with a general liquid fertilizer on a weekly basis, switching to a high potash (tomato) feed when established to help the fruit to form. Keep well watered, paying particular attention when the flowers form. Avoid cold water from the tap. Mulch to prevent evaporation and keep weeded.

Harvesting

Pick as the tomatoes ripen. Gather in the whole outdoor crop before the frosts. Green tomatoes can be ripened by hanging up the trusses in a warm airy spot, by putting on a sunny windowsill or in a paper bag with a ripe banana.

Problems

Outdoor tomatoes are usually trouble free. They can get potato blight, eelworm and mosaic virus.

Recommended varieties

******'Alicante' (AGM)* – A great favourite. Prolific and reliable.
'Beefsteak' – Strong and tall-growing plants with hefty fruits.
**'Golden Sunrise' (AGM)* – Late with small yellow fruits.
'Outdoor Girl' (AGM) – Early with a good flavour.
'Ponderosa Pink' (AGM) – A beef tomato with delicious, rosy red fruits.
**'Shirley' F1 (AGM)* – Quick to mature. Recommended for the organic grower as it is resistant to tobacco mosaic virus, leaf mould and fusarium. It also copes with cooler temperatures than most. Top quality fruits.
**'Tigerella' (AGM)* – Medium-sized tomatoes with decoratively striped fruits.
**'Tornado' F1 (AGM)* – A generous cropper. Large fruit with good flavour.
'Yellow Perfection' F1 (AGM) – A popular old variety; productive, juicy and sweet. A five-star tomato.

CHERRY TOMATOES
**'Gardener's Delight' (AGM)* – One of the most popular tomatoes ever. It never stops producing trusses of sweet cherry tomatoes.
'Sun Baby' (AGM) – Yellow fruits. Prolific.
**'Sweet Million' F1 (AGM)* – Non-stop, tiny, sweet, red fruits.

Seed to harvest: 7–12 weeks.

AUBERGINE

(EGGPLANT)
Solanum melongena
Another member of the deadly nightshade family, the aubergine has been grown in China and India from the 5th century BC. It would seem that the Moors brought them to Spain in the 4–7th centuries AD. Aubergines were believed to cure the wind and the original Sanskrit name *vatin gana*, translated into Arabic *al-badinjan*, meant just that. The Spanish corrupted the name to *alberengena* and the French to

aubergine. The British called it an eggplant as the first types grown here were white and egg-shaped.

Aubergines are grown in the same way as tomatoes, though they are even more sensitive to cold and take longer to grow. Their ideal ripening temperature is 25–30°C (77–86°F). It is possible to grow them outside in hot summers in the south if bought as young plants or started off in the warmth of home or greenhouse. If you just want one or two plants for fun, grow them in pots. They are treated as annuals in the UK.

Aubergine types

Aubergines vary from the big purple ones through to the little white ones the size of a hen's egg. They come in pink, pink flecked, violet, green, orange and striped. The Chinese varieties are long and thin. Breeding has produced some fast-growing cultivars to optimize our chances, and some small types ideal for containers.

Cooking

The aubergine is the base of many famous Mediterranean dishes. These include the *ratatouille* of Provençe, a rich stew of aubergine with tomatoes, garlic, courgettes, onions and peppers; moussaka, the Greek dish of aubergine with minced meat and bechamel sauce; and the Turkish *imam bayeldi* – which translates as the fainting Imam. Presumably the Imam fainted from sheer delight at this dish of aubergine, raisins and tomatoes. The Japanese fry them in tempura batter and serve them with miso sauce. They can be spiced and curried in every possible way.

Soil and situation

The soil should be both well drained and well manured. The situation should be as near tropical as possible.

Sowing and planting

Soak the seed overnight before sowing in spring in a propagator set at 20°C (68°F). Pot on individually when there are three true leaves. Warm the soil for two weeks before planting them out under a cloche, when the minimum soil temperature at night is above 15°C (59°F). Space them 60 cm (24 in) apart. Taller varieties will need a stake.

Cultivation

For sizeable fruits, restrict each plant to five buds by nipping off any in excess of this. Give a general liquid feed weekly, switching to a high potash (tomato feed) when the fruits are forming. Keep well watered and mulched.

Harvesting

Harvest while the fruits are still shiny. When they dull down they become bitter. They will keep for up to two weeks in the fridge.

Problems

Aphids, whitefly and red spider mite in the cold frame. Botrytis in wet weather.

Recommended varieties

'Adona' F1 (AGM) – Big shiny black fruits. Prolific.

**'Black Beauty' F1* – Good old stager with big, purple fruits. High yields.

'Easter Egg' – A modern variety. A quick grower with small white fruits.

'Galine' F1 (AGM) – Big shiny fruits.

'Mohican' F1 (AGM) – Compact with white fruits.

'Moneymaker' – An early variety bred to cope with slightly cooler conditions. Good flavour. Recommended by GW.

'Rima' (AGM) – Suitable for growing outside under cover. Good flavour. Purple fruits.

'Vernal' (AGM) – Good for cold fame or cloche. Good-quality, vigorous plant.

**'Violetta di Firenze'* – A pretty violet version, occasionally striped. Sensitive to cold.

Seed to harvest: 16–24 weeks.

PEPPERS AND CHILLIES

Capsicum annuum

Sweet peppers, bell peppers and chilli peppers are native to South and Central America. Seed has been found dating back to 7,000 BC. They were known in Europe long before Christopher Columbus brought them back on his second voyage to the West Indies in 1493, mistaking them for the black table pepper, *Piper nigrum*.

Another member of the *Solanaceae* family, chillies grow fast. The hotter the weather, the more fiery they become. They can be grown successfully in the south in the UK in the same way as outdoor tomatoes. You double your chances with the new F1 hybrids.

Pepper and chilli types

Sweet peppers, pimento and bell peppers are annuals and the easier of the two types to grow in Europe. There is a great variety of shapes – bell- and box-shaped, bonnet-shaped or long and thin. Colours vary from yellow to near black. The common green ones take on their colours – red, yellow or purple, when they are fully ripened.

Chilli peppers are perennials. The fruits are usually smaller than sweet peppers and fierier. The heat of chilli peppers is measured in units of alkaloid capsaicin. Jalapeños measure in at 3,000–6,000 units, whereas a habanero can reach (a near fatal) 500,000. Most bear red fruits, though there are yellow and white ones too.

Cooking

Recipes with chilli pepper appeared in British cookbooks from the 18th century. The Victorians greatly enjoyed spicy Indian dishes. The Spanish dry and crush them and use the resulting *pimienta* as a condiment. The hotter chillies are used in Mexican salsa, Indian vindaloos and other curries. Chilli powder is usually made from the Mexican *chile ancho* with added cayenne. Indian, Schezuan and Thai cuisine make much use of the dried chilli pepper made from the cayenne types.

Soil and situation

Peppers and chillies like a fertile, free-draining soil in full sun.

Sowing and planting

Sow seed under cover in March or April, potting on or transplanting into growing bags, when there are three true leaves. Keep the plants warm until they can go outside under cloches when night temperatures are over 16°C (61° F). Ideally, this should be when the first flowers are forming. They will need a stake. Feed weekly and keep well watered.

Harvesting

Pick the fruits while still shiny and green to encourage more to come.

Problems

Whitefly, red spider mite and aphids in the cold frame. Botrytis in wet weather.

Recommended varieties

SWEET PEPPERS

'Ariane' F1 (AGM) – Prolific yellow fruits.

**'Canape' F1 (AGM)* – Bred for cooler climates. Red fruits. Good yields. Ready in 60 days.

**'Gypsy' F1 (AGM)* – Early, prolific and popular. Pale green, ripening to bright red, pointed fruits.

'Lipstick' – Productive and easy. Green fruits turning to red.

'Luteus' F1 (AGM) – With yellow fruits.

'Mavras' F1 (AGM) – Handsome black fruits

'New Ace' F1 (AGM) – Tolerates cooler climates. Red fruits and high yields.

CHILLI PEPPERS

'Apache' – Hot red, round fruits.

'Jalapeño' - A fiery ingredient in Mexican cooking. Dark green, changing to red.

'Ring of Fire' – Cayenne type. As it turns from green to red, it lives up to its name.

'Tabasco Habanero' – This will blow your head off! Used in West Indian sauces.

MINI VARIETIES

Sweet peppers and chilli peppers are ideal for container growing. *'Ace' F1 (AGM)*, *'Vidi'* and *'Bell Boy' F1 (AGM)*, are compact varieties of sweet peppers.

Seed to harvest: 20–28 weeks.

SWEETCORN

Zea mays

Sweetcorn was developed from maize – a giant domesticated grass. It is the third most important grain crop worldwide after wheat and rice. Maize seed has been found in prehistoric settlements in Mexico. It was introduced to Europe during the Spanish Conquest. The USA (the Corn Belt), China, Brazil and Mexico lead the field in maize growing for grain and animal fodder.

Sweetcorn types

Modern American breeding has developed sweeter types of corn, known as the supersweet and extra sweet varieties. They have also worked on faster maturing times, making it a practical proposition to grow sweetcorn in temperate climates. There are many varieties – white, yellow, gold and orange. There is also black sweetcorn, ones with kernels of different colours, plus varieties that can be eaten whole when young. There are early, mid-season and late varieties which can be sown a week or two apart to give you a longer season.

Cooking

If freshly picked before the sugar turns to starch, sweetcorn is memorably delicious. It is particularly good barbequed (with husks on) and eaten on the cob. The Americans are fond of corn fritters and sweetcorn chowder. In Argentina, sweetcorn is puréed with milk, eggs, sweet peppers and tomatoes to make *humitas* – a side dish or dip. In India, they spice up the kernels and eat them in chapatis. The small young kernels are used whole in oriental stir-fries. If the kernels are left to harden, they can be cooked along with another variety of maize – popcorn. Sweetcorn doesn't pop but expands to twice its size.

Soil and situation

Choose a warm and sheltered site with well-drained but moisture-retentive, light soil on the acid side, pH5.5–7.

Sowing and planting

Sow seed under cover in April at 20°C (68°F) in modules, root trainers or tubes as sweetcorn resents disturbance. Plant out when the seedlings are 7.5 cm (3 in) high, spacing them 35 cm (14 in) apart in rows 90 cm (3 ft) apart. Sweetcorn is wind-pollinated and is best planted in a square block Covering with a frame or planting through polythene will provide extra warmth if necessary.

Alternatively, wait until May or June and sow outside on pre-warmed soil that has reached a minimum of 16°C (61°F).

Cultivation

Corn has shallow roots so weed carefully by hand, not with a hoe. Keep mulched. Earth up the plants as they grow for extra root anchorage. A good watering when in flower and when the kernels are swelling will improve the crop. You may need to stake the plants if the site is at all windy.

Harvesting

When the tassels, or silks, turn brown, the corn is ripe. Test further by pushing a fingernail into a kernel. The juice should be milky. Twist off the cobs with one hand while supporting the stem with the other.

Problems

Mice and birds.

Recommended varieties

'Conquest' – Supersweet. Early and better than most on cold soil.

★'Dickson' (AGM) – Supersweet. Early maturing, tall plants. Highly recommended.

'Dynasty' (AGM) – Supersweet. Mid-season variety. Tall plants with yellow cobs.

'Honey and Cream', *'Peaches and Cream'* and *★'Honey Bantam'* have white and yellow seeds.

'Kelveden Sweetheart' – Fast-maturing variety with long, well-filled cobs.

'Start-up' (AGM) – Supersweet. Fast-maturing variety. Fine quality cobs.

'Sundance' F1 (AGM) – Early to mid-season. Stocky habit. Good vigorous variety for the British climate, as is *'Summer Flavour'*.

MINI VARIETIES

'Minisweet' F1 – Miniature variety bred for sweetness.

★'Sweet Nugget' F1 – As above.

Seed to harvest: variable, but around 120 days.

OKRA

(LADY'S FINGERS, BHINDI AND GUMBO)

Hibiscus esculentus

Okra is a tropical climbing annual with beautiful large yellow flowers (like those of its relative, the hibiscus), hairy leaves and edible seedpods. Okra is thought to have originated in Ethiopia. It was cultivated in Egypt in the 12th century BC. It is grown widely in Africa, India and the Orient. As it is a tropical plant, attempting to grow it outside in Britain is taking a calculated risk.

Okra types

Heights vary from 90 cm (3 ft) to 3.6 m (12 ft). The seedpods are smooth or ridged, usually green but there are white and red varieties and others with green pods and red stalks.

Cooking

Okra is an integral part of the Creole cuisine of the Southern States, particularly Louisiana. It is the base ingredient of the fish, meat or vegetable gumbos served with rice. Okra's mucilaginous juices thicken the stews. The dried seeds are ground for a coffee substitute. In the Mediterranean, okra is classically stewed with tomatoes and garlic, dressed with lemon and parsley. Okra is much used in India and Pakistan for vegetable curries and fritters.

Soil and situation

Okra likes a rich, fertile and well-drained soil in a suntrap position.

Sowing and planting

Sow seed indoors about six weeks before the end of the frosts, mid-April in the south. The seeds have a tough coat so soak them for 24 hours before sowing. Sow in a propagator set to 16°C (61°F). Transplant into individual pots when they are large enough to handle.

Harden off (carefully) when about 10 cm (4 in) tall. When the outdoor soil temperature has reached 16°C (61°F) it should be safe to transplant them, 45 cm (18 in) apart. If in doubt, cover with fleece or polythene. Alternatively, let the nurserymen take the strain and buy a small plant in June.

Cultivation

Stake or tie in the shoots. Keep well watered and weeded. Feed weekly with a general liquid fertilizer and a high potash (tomato) feed when the fruits are forming. Pinch out the growing tips from time to time to encourage side shoots.

Harvesting

Pick the pods when about 7.5 cm (3 in) long on a regular basis as they quickly become stringy and coarse.

Problems

Aphids, red spider mite and whitefly in the cold frame.

Recommended varieties

'Burgundy' – Red stems, green leaves. Said to cope better than most with our climate. Productive, tender fruits. Quite tall, up to 1.2 m (4 ft).
'Clemson's Spineless' – Introduced in 1939 and still a popular choice.
**'Pure Luck' F1* – Vigorous and fast. Recommended by seedsmen for the UK.

Seed to harvest: Four to five months.

STEM VEGETABLES

Celery, celeriac and Florence fennel are grown for their edible stems. Celery and Florence fennel can be difficult to cultivate as they will bolt if conditions are not perfect. They all need fertile, well drained but moisture-retentive soil with plenty of organic matter.

CELERY

Apium graveolens var. *dulce*
Wild celery is a Mediterranean marsh plant thriving in damp ditches, often near the coast, in many parts of Europe. It was widely used as a medicinal plant by the Greeks and Romans. Homer mentioned it in *The Odyssey*. Known as smallage in England, it was used to flavour soups. The modern equivalent is cutting celery. The celery as we know it today and celeriac (which is grown for its bulbous lower part of the stem) were bred in the 17th and 18th centuries respectively.

Growing celery from seed represents a challenge, even for the most experienced gardener. If you are not seeking to test yourself, I would suggest buying small plants for growing on. The main problem with celery is its tendency to panic – to flower and die if things are not going smoothly. Being a biennial, it should flower in the second year. However, if it thinks winter is coming and the temperature drops below 10°C (50°F) for more than a few hours, it will probably flower in the first year and all your efforts will be lost. For this reason it is vital to delude it that it is summer right from the start. Another problem is that slugs are drawn to celery like a magnet.

Celery types

Self-blanching or summer celery These new cultivars, developed in the USA, cut out the work of growing celery in trenches and earthing it up. They are not frost-hardy however. The stalks are bred to be pale yellow – sometimes with a pink flush, as well as in various greens.
Trench celery or winter celery This is the traditional British celery and has the best flavour. It can take a mild frost and some varieties can stand out in winter. There are the hardier pink- and red-stemmed types as well as green ones and large and small varieties.

Cooking

The inner stalks of celery are

delicious eaten raw. In the UK, celery stalks are a classic accompaniment to cheese. In France they eat them as an appetizer with sea salt. They are the base ingredient of Waldorf salad. Thin sticks can be stir-fried. They can be stewed or braised as a side vegetable or made into soup. The tougher outer stems and winter celery are useful for flavouring stocks and stews. Celery is a good source of vitamins A and C, iron and fibre, especially when eaten raw.

Soil and situation

Celery needs to grow fast in deep well-drained soil of pH6.5–7 with plenty of organic matter and water. It thrives best in the peaty Fens.

Sowing and planting

Self-blanching or summer celery The best method is to sow treated seed in modules in mid-spring in a propagator set to 16°C (61°F). Celery seed is finer than dust, so mix it with sand or vermiculite to make it more manageable. It is one of the few plants that germinates in the light, so scatter seed as thinly as possible on top of the dampened compost. When the seedlings are big enough to handle, thin out to one per module and transfer them into boxes with glass on top, keeping them at about 13–16°C (55–61°F). When they have five or six true leaves in early summer, they can be hardened off and transplanted outside in May or June. If it's still cold when they are ready to go out, trim them down and keep them in the warm a little longer.

Plant them with the base, or crown, at soil level. To get perfectly blanched plants, arrange them close – 20 cm (8 in) apart – in a square block so they shade each other. Cover with fleece or cloches for a few weeks and protect against slugs. If you put collars (strips of cardboard or newspaper, covered in black plastic and tied with string) around the outer ones, they too will come out perfectly blanched. Take off the covers from time to time to check for slugs.

Trench celery or winter celery Trench celery is sown in the same way and at the same time as the self-blanching types. The difference comes in the cultivation. The winter before you plan to grow celery, dig a trench about 30 cm (12 in) deep and wide and partially fill it with well-rotted manure or compost. The remaining soil is left on the side for earthing up. The young plants are set about 30 cm (12 in) apart.

Let them grow about 30 cm (12 in) high then tie the tops loosely together with soft string and begin to earth up. Water well before adding soil at the rate of 7.5 cm (3 in) every three weeks. Take care to avoid getting soil into the hearts, as it could cause them to rot, and don't pack it around the plants.

A more modern approach is to grow them on the flat with collars. When the plants are about 30 cm (12 in) high, place loose collars of cardboard or brown paper around them, tied with string. Cover this with polythene to prevent the paper disintegrating in the rain. Small sections of plastic pipe or roofing felt are also effective. The key is to always leave about a third of the plant out in the open. Lift the covers from time to time to check for slugs.

Cultivation

Celery needs a constant supply of water and a weekly liquid feed (seaweed is good) from early summer onwards. When the frosts come, cover with straw and stop feeding.

Harvesting

Self-blanching or summer celery Test for readiness by snapping off an inner stalk. They should be ripe for harvesting between the end of July and September. It's important to lift them before they become stringy and tough. You can cut off the outer stalks as you want them or water well before lifting the entire plants. Clear the crops before the frosts set in. They will keep for several weeks in a cool shed or in the fridge.

Trench celery or winter celery Trench celery is ready in autumn and early winter. If you have grown it in the trench method, carefully remove the soil that you used for earthing up before lifting the plant with a spade. Wash off and remove unnecessary leaves before storing upright in boxes with dampened sand at the roots. Store in a cool but frost-free shed.

Problems

Celery crown rot, celery leaf miner, carrot root fly, celery fly, celery leaf spot, slugs and snails, and calcium deficiency.

Recommended varieties

SUMMER CELERY

**'Celebrity' (AGM)* – Self-blanching. Early, nutty and crisp with bolting resistance.

'Ivory Tower' (AGM) – A variety that is nearly stringless. Tall with good flavour.

'Lathom Self Blanching' (AGM) – Hearty and well flavoured, a vigorous early variety. A popular choice.

'Pink Champagne' – An ornamental pink variety.

'Tango' (AGM) and **'Victoria' F1 (AGM)* – Both are reliable green types with good flavour.

TRENCHING CELERY

'Giant Pink – Mammouth Pink' (AGM) – Green with a pink tinge. Harvest early winter.

'Giant Red' – Known for its hardiness.

'Solid Pink' – Introduced in 1894. A good, hardy variety.

Seed to harvest: 11–16 weeks.

CELERIAC

Apium graveolens var. *rapaceum*

Though a close relative to celery, celeriac is a much easier bet, being hardier and less temperamental. It can stay in the ground with cloches or a covering of straw right through winter. It does, however, take six weeks longer to mature than celery and you may not achieve the size you find in the shops.

Celeriac types

Not many. Newer cultivars aim for a smooth skin that is easier to peel.

Cooking

Celeriac makes a tasty salad, popular on the Continent. The swollen stem is blanched before being cut into julienne strips and mixed with mayonnaise. As a side vegetable, it can be cooked for about half an hour, then cut into chips and sautéed. Alternatively, boil and mash it. It can also be topped with cheese melted under the grill. In Eastern Europe, it is made into soup with wild fungi. The leaves can be used for flavouring.

Soil and situation

It needs a rich moist soil, high in organic matter. It prefers an open situation but can cope with a little shade. It's an ideal vegetable for a damp patch.

Sowing and planting

Start the seed off in a propagator at 18°C (64°F) in mid-spring or outside under glass in late spring. Harden off before planting outside after the last frosts. If the temperature drops, delay by clipping off the tops. Plant so the crown is just above soil level, leaving about 30 cm (12 in) between each one.

Mulch around them and keep up the watering. Give them some liquid seaweed every couple of weeks. In midsummer, remove side shoots and if there is more than one growing bud, remove it or them. Snip off the outer leaves so the sun can get to the crown and ripen it.

Harvesting

Celeriac should be ready by autumn. In mild parts of the country it can be left in the ground covered with straw through winter and harvested as needed. Alternatively it can be dug up and stored in damp sand in a cool shed. Remove most of the leaves but don't take out the central tuft as the root will waste away trying to produce more leaves.

Problems

Apart from slugs, celeriac is generally trouble-free though it can get the same pests and diseases as celery.

Recommended varieties

'Diamant' (AGM) – Medium-sized bulbs with clean white skin.
'Ibis' (AGM) – Pleasing, small round balls.
'Kojak' (AGM) – Smooth, flattened shape.
'Marble Ball' – A good storer.
**'Monarch' (AGM)* – Succulent, firm, white flesh and smooth skins. Harvest in autumn or early winter. A popular choice.
'Prinz' (AGM) – An early variety with big bulbs which is slow to bolt. Reputed to be good for early protected crops or outdoor production.
'Snow White' – Big yields. The flesh doesn't discolour easily.

Seed to harvest: 26 weeks.

FLORENCE FENNEL

Foeniculum vulgare var. *azoricum*

Florence fennel, with its sweet aniseed taste and elegant feathery foliage, thrives in the Mediterranean climate. A pretty plant, unfortunately it is not the easiest to grow. Like celery it is inclined to bolt at the slightest difficulty – if it is too cold or too hot, doesn't have enough water or if it's disturbed for transplanting. If it does go to seed, one good thing is that you can use the leaves for flavouring.

Florence fennel types

Not many. Some grow faster than others and some have bolt-resistance.

Cooking

Fennel tastes of aniseed and is sometimes partnered with Pernod. The heart of it can be eaten raw in the same way as celery though it's a tougher proposition. In the south of France it is served *à la Niçoise*. The fennel is boiled, covered with tomato sauce (with olives and wine added) and baked in the oven with a cheese topping. Another French method is *à la Grecque*. The fennel is stewed in a tomato sauce and served cold as a salad. Fennel is a classic accompaniment to fish or chicken, parboiled and roasted with garlic.

Soil and situation

It does best on a light, well-drained soil rich in organic matter. It prefers sandy soils.

Sowing and planting

Buy fast-growing, bolt-resistant seed. To improve your chances further, sow outside to avoid transplanting shock in May when the soil is at least 10°C (50°F). Space them 30 cm (12 in) apart each way. Sow more in late summer for an autumn crop. Use fleece at each end of the season.

Cultivation

Keep very well watered and spread mulch around the plants. When the bulbs start to form, either earth up to half way up the bulb or tie card-board collars around them to blanch the stems. Cover with fleece as the nights draw in.

Harvesting

Cut the bulbs off, leaving the roots in the ground to throw out a few shoots. Sometimes the bulbs don't swell at all but you can still eat the rest of the plant. The leaves are a classic accompaniment to fish.

Problems

Bolting, slugs.

Recommended varieties

'Carmo' F1 (AGM) – Fine quality and fast maturing.
'Dover' (AGM) – An early variety.
'Heracles' (AGM) – Fast to mature. The RHS describes it as "excellent quality".
'Romanesco' – Resistant to bolting. Hefty round bulbs up to 1 kg (2 lb) in weight.
**'Zefo Fino' (AGM)* – Good bolt-resistance. Medium-sized, well filled white bulbs. Very ornamental. The RHS describes it as "excellent quality".
'Zefo Tardo' – A bolt-resistant, late variety with solid bulbs.

Seed to harvest: 10–15 weeks.

SALAD LEAVES

Home-grown salad leaves should save the average household a small fortune, judging by the amount they cost in the shops and how many we eat. Lettuce is, without doubt, the most popular salad vegetable worldwide. It was recorded by Herodotus as being served to the Persian Kings in the 6th century BC, and was greatly enjoyed by the Romans who ate dozens of different types. Add to these the other fancy salad leaves – chicory, corn salad, endive, lettuce, mustard and cress, rocket, summer and winter purslane, Good King Henry, American land cress, the cut-and-come-again Oriental brassicas and the spinach-type greens – amaranthus and orache – and you could be producing an endless supply of decorative and nutritious salad material all year round.

Salad leaves, being fast and low growing, are ideal for intercropping, undercropping and catch cropping. If you sow a few seeds every couple of weeks in the growing season you can have non-stop salads. If you choose suitable cultivars you can keep endive, chicory, corn salad and rocket going through winter with protection or through forcing. In fact you could easily start up a stall to raise funds for the allotment at the local farmer's market as some enterprising people already do.

Lettuce and most of the salad leaves are prone to aphids, cutworms, slugs and snails, aphids and tip burn.

Although salad leaves can be thrown into delicious soups like the Italian zuppa di verdura that is comprised of everything green to hand and in season, the prime use of salad leaves is for just that. Always tear rather than cut the leaves to avoid bruising. Lettuce is a good source of vitamin A and a moderate one of vitamin C.

LETTUCE

Lactuca sativa

The name lettuce is derived from the Old French *laitue* meaning milky, referring to the milky substance that comes out of the leaves. The cultivated lettuce evolved from the wild forms that originated in Asia Minor, Iran and Turkistan.

Lettuce types

There are four main types – butterhead, crisphead, Romaine (or Cos), and looseleaf. The looseleaf types are much easier to grow than the hearting types. As they don't store well, they are expensive to buy and really worth growing yourself. Small lettuces like Little Gem are also easy. Crisphead lettuces (like Iceberg) and the butterheads are the most demanding, needing quantities of water and fertilizer.

Modern breeding has put in bolt- and disease-resistance and come up with many interesting variations of colour and shape for all types. You can also buy mixtures of various lettuce types or cut-and-come-again mixed seedlings. Known as saladini in the UK, *mesclun* in France and *misticanza* in Italy, they are fun to try out to see which varieties you like best. Alternatively you could exchange seed with neighbours and make your own mix.

Cooking

Lettuce is normally eaten raw as a prime salad ingredient. The French make very good lettuce soup and also put a few leaves in with young peas to make *petits pois a la Française*. If you have a glut, try braising them.

Soil and situation

The ideal is light sandy, fertile, moisture-retentive soil with free drainage. Soil which has been fertilized for a previous crop is ideal. If your soil is heavy and wet, it is best to raise the beds. Lettuces like an open, sunny situation early in the season and part shade during the hot months.

Sowing and planting

Early and late lettuces are usually sown under cover and the mid-season ones sown where they are to grow. There are just a few points to watch. Leaf salads are sensitive to temperature. The ideal for germination is within the range of 10–20°C (50–68°F). Any lower and germination will be erratic and over 25°C (77°F) the seed may go into dormancy. High temperatures can make established plants bolt or become coarse and unappetizing.

If sowing indoors, use biodegradable modules as lettuces can suffer from transplanting shock. Thin when about 5 cm (2 in) high. Move them when they have six leaves, taking care to plant them so the leaves are just clear of the ground – any lower and they may rot off, any higher and they may not grow to their full potential. Hardy lettuces planted in autumn will stand out through winter (under cloches or in the cold frame) in mild areas. In colder places they can be sown in late winter (February) for spring eating.

Cultivation

Cover early sowings with cloches or fleece. Keep well watered in summer.

Harvesting

Snip away at looseleaf types as you want them. If you cut the lettuce right off it may resprout from the base two or three times. Cut headed lettuce when they feel solid and hearty. If you leave the root in they may also resprout to produce a few extra leaves.

Recommended varieties

BUTTERHEADS FOR SUMMER

'Enya' (AGM) – Good as a good garden variety. Well hearted and tidy.

**'Lillian' (AGM)* – Solid and round, well filled.

**'Merveille de Quatre Saisons'* – An old variety. Dark bronze leaves. Popular in France.

'Roxy' (AGM) – Bronze tips to green leaves. Uniform.

**'Webb's Wonderful'* – Introduced in 1890. Highly popular and reliable.

BUTTERHEADS FOR AUTUMN UNDER COVER

'Avondefiance' (AGM) – Quite large, mid-green heads.

'Clarion' (AGM) – Pale green leaves.

'Sunny' (AGM) – Early and quick to mature.

CRISPHEADS

'Beatrice' (AGM) – Small, quick growing and early.

'Iceberg' – Very crisp and tender.

'Lakeland' (AGM) – Quite late and resistant to root aphids. Solid and hearty.

'Sioux' (AGM) – A solid lettuce with a green heart with red tips on the outer leaves.

ROMAINE OR COS

**'Bath Cos'* – Introduced in 1880. Massive heads.

'Corsair' (AGM) – Vigorous.

'Lobjoit's Green Cos' (AGM) – A delicious old variety with smooth mid-green leaves. Spring and summer.

'Pinnokkio' (AGM) – Disease-tolerant. Crisp and sweet tasting. Medium size and dark green.

'Winter Density' – A first-class candidate for winter growing under a cloche or coldframe. Dark green with a good heart. Medium size.

LOOSELEAF LETTUCES

'Delicato' (AGM) – Similar to 'Red Salad Bowl' but with bright red leaves.

**'Lollo Rossa' (AGM)* – Green with a bronze frill. **'Lollo Blonda'* is the pale green version.

**'Oak Leaf'* – An old variety, aptly named. Green with brown tints.

'Red Ruffles' (AGM) – Similar to 'Red Salad Bowl' with blistered leaves.

**'Red Salad Bowl' (AGM)* – Big oakleaf type, with bronzy red tints.

**'Salad Bowl' (AGM)* – A pretty, frilly, light green lettuce.

MINI VARIETIES

**'Blush' (AGM)* – Baby iceberg with pink tinges.

**'Frillice' (AGM)* – A small, dark green, frilly looseleaf lettuce.

**'Little Gem'* – Small but delicious Cos type, a good candidate for early sowings.

'Minigreen' (AGM) – A neat little crisphead, just right for one person.

Seed to harvest: 4–14 weeks, depending on the season and the variety.

ENDIVE AND CHICORY

Endive and chicory are easily confused. This is not helped by the fact that the French call endive chicory and chicory endive. They are closely related to each other and to the dandelion – hence their bitterness. Growing wild in parts of Europe and the Middle East, they are one of the oldest of cultivated plants. Some varieties were known to the Ancient Egyptians. By the 17th century they were widely grown in British gardens.

ENDIVE

Cichorium endivia

Endives are trouble-free plants, more tolerant of heat than lettuce, and they make a lively addition to the salad bowl. The new varieties are less bitter than the old. They can be grown to maturity or used for cut-and-come-again.

Endive types

Endive comes with frilly leaves (frisée) and as a broad-leaved plant (Batavian, endive or escarole). The frisées are best for summer and autumn. They can survive a light frost while the Batavians, given cloche protection, can go well into winter.

Cooking

Generally used for salads, though they can be blanched and eaten hot. Jane Grigson recommends a lusty Italian dish, *cicorie rosse de Treviso ripiene*. The heads are stuffed with olives, anchovy, pine kernels, herbs, garlic and wine.

Soil and situation

The best soil for them is light and free draining. It should be reasonably fertile, though too much nitrogen will encourage lush growth and pests They enjoy an open situation and like a little shade in the heat of summer. Give them a sheltered spot if you are growing them into winter.

Sowing and planting

They can be sown outside or in, at a temperature of around 22°C (72°F). Sow thinly about 1 cm (1/2 in) deep. Sow from June to July for autumn crops and hardier types in August for winter crops. Sow successively for cut-and-come-again.

Cultivation

Keep weed-free. Water in dry weather and mulch. To remove bitterness, blanch for two or three weeks before they are ready to eat. Tie the leaves together with soft string or raffia, or cover them with a bucket. If you just want to blanch the centre, use an old plate. Make sure they are completely dry first and that there are no hidden slugs lurking in the leaves, anticipating a private banquet.

Harvesting

Pick individual leaves or cut the head off and leave the plant to resprout.

Problems

Slugs, lettuce root aphid, caterpillars and tip burn.

Recommended varieties

***'En Cornet de Bordeaux'* – A Batavian type. A hardy old French variety for cut-and-come-again through winter.

'Grosse Pancalieri' – Self-blanching variety – curly with rosy midribs. March to September.

'Jeti' (AGM) – Attractive bright green plants with a curly leaf.

Seed to harvest: 12 weeks.

CHICORY

Cichorium intybus

The chicories provide good bitter leaves to perk up cool weather salads. The long-rooted types have been used roasted and ground as a coffee substitute since the Napoleonic Wars.

Chicory types

They fall into three groups.

Witloof or Belgian chicory Witloof is Flemish for white leaf. It is said that a Belgian farmer by the name of Jan Lammers returned from the wars in 1830 to discover that the chicory he had forgotten in his cellar had grown sweet and plump in the dark, warm conditions. Ever since it has been grown for forcing in autumn for winter eating.

Radiccio or red chicory This is a decorative, hearting chicory that flushes red for autumn eating. New cultivars have been bred to colour early and have more heart.

Sugarloaf chicory These non-forcing types are like big hearty Cos lettuces. The outer leaves are discarded and the heart, which is naturally blanched and mild tasting, is eaten – hence the name.

Cooking

All the chicories are eaten raw as an interestingly bitter and decorative salad ingredient. They can be braised as a side vegetable.

Soil and situation

They like a fertile and free-draining soil – not too rich as too much manure can cause the roots to fork. Choose a sunny site, though a little shade won't harm.

Sowing and planting

Witloof Sow sparingly outside in May or June about 1 cm (1/2 in) deep. Water until the seeds come up, then leave them fairly dry to encourage root growth. Thin to 20 cm (8 in) apart when the first true leaves appear.

Radiccio The traditional and the easiest time to sow radiccio (or red chicory) is in midsummer for autumn eating. It is broadcast where it is to grow and thinned to about 30 cm (12 in) apart. For a summer crop it can be started off in modules under cover. For the best of both worlds use some for cut-and-come-again and leave others to hearten up. Some varieties can withstand light frost.

Sugarloaf chicory Sugarloaf chicories are sown in mid-spring indoors, or outside at any time through summer when the temperature is above 10°C (50°F). For a mature crop, sow outside in midsummer for autumn eating. Sow a few seeds every two weeks for a cut-and-come-again crop.

Cultivation

Witloof Keep weed-free and water in dry weather. In mild areas you can blanch the plants outside. In autumn, cut off the leaves down to 5 cm (2 in). Cover the whole plant with soil, straw or leafmould and put a bucket on top.

Alternatively, blanch them indoors. Dig them up in late autumn or early winter, cut the leaves off as before and trim the roots to 30 cm (12 in). Store them in moist sand in a frost-free shed. When it suits you, plant a few in large pots of moist soil with the crowns just showing. Keep in the dark at a temperature no cooler than 10°C (50°F) and the chicons will grow within a month. If they are broken off carefully, more will form.

Sugarloaf chicory If covered with cloches or fleece in autumn, these plants will continue to grow for some time. You can extend the season further by lifting them, and growing them on in pots indoors. They don't go over quickly like lettuce but will keep in the ground for several weeks.

Harvesting

Witloof When the chicons have formed, break them off carefully. With luck, more will form.

Sugarloaf chicory Keep picking young tender leaves for salads. Headed chicory is usually harvested by autumn. if you leave the stump in the ground it will resprout. Alternatively, leave them to grow on under cloches. You can extend the season this way by several weeks.

Problems

Generally trouble-free, though slugs may be a problem.

Recommended varieties

WITLOOF

**'Zuckerhut – Witloof de Brussels' (AGM)* – A great traditional variety.

SUGARLOAF CHICORY

'Pan di Zucchero' (AGM) – Dark leaves but blanches well.

**'Sugarloaf'* – Resistant to cold, a good choice for winter under cloches. A big-hearted chicory with a mild flavour.

RADICCHIO

'Leonardo' (AGM) – Good and hearty red variety.

**'Palla Rossa' (AGM)* – Dark red with white veins.

Seed to harvest: variable.

CORN SALAD

(LAMB'S LETTUCE, MÂCHE)

Valerianella locusta, V. eriocarpa

Known as lamb's lettuce because it grows wild in the lambing season, and as corn salad as it is often seen in corn-fields, this is another gourmet green leaf for winter. It grows almost as easily as a weed and will self-sow freely if allowed. Provided it is kept covered in cold weather, it will carry on through winter. A real winner.

Corn salad types

The French type is compact with a small leaf, while the English or Dutch ones have longer leaves.

Cooking

A nice addition to salads. The French type picked whole looks particularly decorative.

Soil and situation

Not fussy. The ideal is deep, fertile and free-draining soil in sun or part shade.

Sowing and planting

Don't sow before midsummer as it is likely to bolt. Sow a few seeds every couple of weeks after that for a succession. Sow 1 cm (1/2 in) deep, thinning to about 15 cm (6 in) apart when they have three or four true leaves. Give the winter crop some cover.

Cultivation

Keep weed free and well watered in dry spells

Harvesting

Pick as you want for cut-and-come-again crops.

Problems

Generally trouble free. Slugs and snails may be a nuisance.

Recommended varieties

'Cavallo' (AGM) – Small-leaved, neat French type.
'Verte de Cambrai' – Traditional French variety. Small leaves, looks inviting harvested whole in salads.
'Vit' – A modern variety with dark green leaves. Vigorous and good for winter production.

Seed to harvest: 6–12 weeks.

ROCKET

(ROQUETTE, RUCOLA, ITALIAN CRESS)

Eruca sativa, E. versicaria

Rocket, the super trendy salad leaf with a mustard kick, is expensive to buy but incredibly easy to grow.

Rocket types

Not many.

Cooking

Used a lot in Italian cooking, particulary salads, risottos etc.

Soil and situation

Almost any. Light shade is preferred in midsummer.

Sowing and planting

Scatter the seeds, a few at a time, between late spring and autumn. It is usually sown in drills and thinned to 15 cm (6 in) apart. It can be grown under crop protection in winter.

Cultivation

Keep watered in dry weather.

Harvesting

Pick individual leaves or cut off the tops completely. They will resprout several times. Pick the leaves when young – they get hotter as they coarsen.

Problems

Unlikely.

Seed to harvest: about 40 days.

CELTUCE

(CHINESE STEM LETTUCE, ASPARAGUS LETTUCE, WHO SUN)

Lactuca sativa var. *augustana*

Celtuce is an ancient Chinese vegetable introduced to Europe in the 19th century. It was the Americans who gave it the name 'celtuce' as the stems are used like celery and the tops are like lettuce. Celtuce can cope with light frosts and summer temperatures up to about 27°C (80°F). In the UK, it is grown as a summer crop in much the same way as lettuce.

Celtuce types

In China there are white-, purple- and red-leaved varieties, round or pointed in shape, dull or glossy. Sometimes these are included in Oriental seed mixes. In the UK there aren't many named varieties as yet.

Cooking

The stems are thinly sliced for stir-fries. In China they are made into Shanghai pickles. The leaves can be eaten in the same way as lettuce (though they are inclined to be bitter) or cooked as greens. If you want to keep the stem for a few days in the fridge don't cut the top off.

Soil and situation

They prefer very fertile, free-draining soil around pH7. Add plenty of organic matter. Plant in sun in spring, part shade in summer.

Sowing and planting

The easiest method is to sow a few seeds every couple of weeks outside from mid-spring onwards. Sow in drills about 1 cm (1/2 in) deep, thinning to a distance of 10 cm (4 in). If you want to carry on into winter, use crop covers.

Cultivation

Keep well watered and mulched. Feed with liquid fertilizer every two to three weeks.

Harvesting

They are ready for harvesting when the stalk is about 30 cm (12 in) long. Either pull the plant up or cut it off at ground level.

Problems

As for lettuce.

Recommended varieties

'Zulu' – A variety which has been bred for cool climates.

Seed to harvest: 3–4 months.

AMERICAN LAND CRESS

(WINTER CRESS, HERBE DE ST. BARBE)

Barbarea vulgaris

Land cress tastes peppery like watercress, but it doesn't need special conditions as watercress does. It was cultivated widely in England in the 17th century but fell from fashion. A great bonus is that it is an all-year-round crop.

Land cress types

There are no named cultivars.

Cooking

American land cress is especially useful in winter. It is ideal for spicing up salads, making soups or garnishing.

Soil and situation

For best results provide moist, fertile soil. Add plenty of humus if necessary. Position in light shade

Sowing and planting

The easiest way is to sow seed 1 cm (1/2 in) deep outside in July or August in drills, thinning to 15 cm (6 in) apart. This should give you a harvest in autumn and through winter. Give it some protection in winter for more delectable crops. If you leave a few to flower the following spring they will self-seed for a midsummer crop. You could also save the seed for sowing the following summer for cropping the next winter. For summer and autumn crops, sow in early and late spring.

Cultivation

Keep well watered in dry spells. If allowed to dry out they will coarsen.

Harvesting

Pick the outer leaves as you want them as cut-and-come-again.

Problems

Generally trouble-free. Flea beetle.

Seed to harvest: 8–12 weeks.

PERENNIALS

Asparagus, cardoons and globe artichokes will stay in the ground for years. Site them carefully where they will not cast shade on other plants and prepare the soil for a long stay. Once established, perennial vegetables need little attention other than routine care.

GLOBE ARTICHOKE

Cynara scolymus

The globe artichoke is a giant thistle from the sunflower family. It is related to the cardoon that is found growing on the sandy coastlines of North Africa. It has large silvery leaves and sends up an architectural spire reaching 1.5 m (5 ft) and taking up about 90 cm (3 ft) of space. If you don't harvest the edible flower bud, you will get a bright blue thistle flower, irresistible to bees.

It is an ancient cultivated Mediterranean plant, well known to the Greeks and Romans, though there may be confusion between it and the cardoon. It was probably Catherine of Medici who introduced globe artichokes to France in 1466 when she married Henry II, bringing with her an entourage of Italian chefs. It has always been appreciated in England, both as a vegetable and an ornamental. It is no trouble to grow.

Globe artichoke types

The flower buds range from green to purple. Modern breeding has produced ever more succulent varieties, some hardier than others.

Cooking

It is the heart of the flower that is eaten in the more mature artichokes, along with the flesh inside the bracts. They are usually boiled and served hot with butter and lemon, or hollandaise sauce, or cold with vinaigrette. If you are pruning the side shoots to get large heads, the tiny young buds that you prune off can be eaten whole as crudités with olive oil, or cooked and served cold. In the south of France and Italy where artichokes are plentiful and cheap, they make many dishes out of the hearts. They are chopped into salads, mashed into a paste and served on bread, or sliced and fricasséed as a side vegetable. They are also stripped down to the hearts (with a few leaves left on to act as 'walls') and stuffed.

Soil and situation

They like well drained, fertile soil of pH6.5–7. Put in plenty of well rotted organic matter as they will stay put for several years. Globe artichokes will not do well in cold wet soil. Position them in sunshine and out of winds.

Planting

Don't attempt to grow artichokes from seed as they are likely to revert to their spiny and less succulent forebears. It is better to propagate from rooted suckers of known stock. These can be taken off existing plants by dividing them in spring (see page 155) or they can be bought from a nursery. Trim back any awkwardly placed roots. Plant them out 90 cm (3 ft) apart with the crown just below the surface.

Cultivation

Keep an eye on them until they establish. Provide a windbreak if necessary and keep watered. Mulch to keep in moisture. As the flower buds form, feed with liquid seaweed every few weeks. You should get one or two artichokes in the first season but more will come in the following years. If you want a bumper crop in the second year, remove any flower buds in the first. For top-quality artichoke production, divide the plants every three years. Earth-up or cover the crowns with straw in winter in colder parts of the country.

Harvesting

Harvest before the flower buds begin to open, starting with the topmost one. Cut them off with 12.5 cm (5 in) of stem. As the plants mature over the years, they will produce more and more. If you want large heads, remove the side buds.

Problems

Aphids may be a problem.

Recommended varieties

'Green Globe' – A popular variety with big, fleshy, succulent heads.

'Purple Globe' – The purple version of 'Green Globe'. They may need winter protection.

'Romanesco' – A purple variety recommended for flavour.

'Vert de Laon' – Hardy and a good flavour.

'Violetta di Chioggia' – A decorative variety with purple heads. Not very hardy.

Offset to harvest: about a year.

CARDOON

Cynara cardunculus

If the artichoke makes a big statement in the border, the cardoon outstrips it. It is a wonderfully dramatic thistle plant, growing to 1.8 m (6 ft) or more. It is usually grown as an ornamental these days. It is the blanched stems and midribs that are eaten, rather than the buds. Cardoons are hardier and easier to grow than globe artichokes. Offsets are hard to find so they are usually grown from seed, and propagated from your own offsets in later years. Decorative varieties can be found in nurseries selling ornamental plants.

Cardoon types

Not many, as the cardoon has fallen from popularity. There are some named cultivars.

Cooking

The best part of the cardoon is the peeled stems, though the young hearts can be eaten in the same way as globe artichokes. The stalks can be eaten raw when peeled and finely sliced. They are peeled after they have been blanched. In Provençe they are made into the traditional Christmas dish *cardons de Nöel*. The blanched stems are covered in béchamel sauce, topped with gruyère or Parmesan cheese, and browned under the grill.

Soil and situation

As for globe artichokes. Position in full sun.

Sowing and planting

Start them off in March in a propagator set at 13°C (55°F). Sow three seeds per module, thinning to the strongest and planting out after all danger of frost. Alternatively, sow three seeds per station, in situ in mid-spring. Sow them 2.5 cm (1 in) deep and 60 cm (24 in) apart.

Cultivation

Keep well watered in dry weather. Feed with liquid fertilizer weekly. Stake them when they reach about 30 cm (12 in) in height. In late summer or autumn they will be ready for blanching. Make sure the plant is dry before tying the stalks loosely together. Tie collars of newspaper or brown paper covered with black polythene around the group of stems, earth up the base and leave for three or four weeks.

Harvesting

Dig up the entire plant and trim off the leaves and roots. Divide the roots for next year's crop (see page 155).

Problems

Usually trouble free. Aphids may be a problem.

Recommended varieties

**'Gigante di Romagna'* – A good culinary variety.

Seed to harvest: 6 months

SEAKALE

Crambe maritime

Seakale is a fine part of our heritage that has been all but lost. It is a coastal plant found in Northern Europe as far as the Black Sea. Gerard describes it in his *Herball* (1597) as to be found on the 'the bayches and brimmes of the sea, where no earth is seen, but sande and rolling pebble stones'. It was gathered wild, blanched under a pile of sand, and sold in markets long before the Victorian gardeners grew it in their kitchen gardens for the gentry. Seakale is herbaceous, dying completely back in winter. It is easy to grow and will keep cropping well for about six years.

Seakale types

There are not many. Some wild seed and a few cultivars are available.

Cooking

Seakale is eaten like asparagus, steamed or boiled and served with melted butter or hollandaise sauce. The young leaves can be cooked with it or used separately as greens.

Soil and situation

Think beach – deep, well-drained, light, sandy soil with a pH around neutral. Choose a sunny site. As seakale will be in the ground for many years, prepare the land carefully and add some general fertilizer a couple of weeks before you are ready to sow or plant.

Sowing and planting

Seakale is usually grown from offcuts or thongs. If you can get hold of some thongs, rub off all the buds except the strongest and plant about 5 cm (2 in) deep and 45 cm (18 in) apart. If none are to hand, sow seed in spring. Seed will give you a small crop in the first year; thongs a bigger one.

Cultivation

Keep weeded, watered and fed (seaweed would be a natural choice) from time to time. In the autumn cut away all the dying foliage. In February or March, cover the crowns with a forcing pot, a bucket or bin, or a frame covered with black polythene. Alternatively, cover the plants with leafmould or sand.

Harvesting

When the fresh spring shoots are about 20 cm (8 in) long they are ready to harvest. Cut them off with a sharp knife. You should get about three cuts from each plant. After this, remove all the covers and feed well so the plants can build up strength for the next year.

Problems

Generally trouble free. Slugs as always, but a covering of sand should keep them at bay.

Recommended varieties

'Ivory White' – Pale, heritage variety.
**'Lily White'* – Pale, good cropper and fine flavour.

Thongs to harvest: for best crops two years, then every year.

ASPARAGUS

Asparagus officinalis

Asparagus is often described as the food of kings. It is said that the Egyptians cultivated it to offer to their gods. The Romans ate it not only in season, but dried it and also froze it in mountain regions. Louis XIV had special green-houses built so he could eat it all year, while Madame Pompadour took an asparagus concoction as an aphrodisiac.

Asparagus is thought to have originated in the Mediterranean but as it grows wild in many countries in Europe, Asia and Africa near fresh water and the sea, this is not known for certain. It will produce for a good twenty years and it is easy to grow, given the right conditions.

Asparagus types

Nowadays the all-male F1 cultivars are recommended. The female plants are less vigorous and inclined to self-seed all over the place. White asparagus, popular in Italy and France, is grown in the dark by earthing up.

Cooking

Really fresh asparagus is a world apart from any you can buy. There is no better way to cook it than to steam it standing up (so the bases get cooked at the same time as the tips) and to serve with hollandaise sauce. It can also be made into soup, used in stir-fries, chopped into quiches or added to salads.

Soil and situation

Asparagus will grow on almost any soil except heavy clay, acid peat or soil that is waterlogged. However, for the food of kings, it is worth getting optimum conditions from the start. The autumn before, designate a sunny and sheltered spot that hasn't been used for asparagus or potatoes. Dig a trench about a spit deep and remove every trace of perennial weeds. The traditional bed is 1.2 m (4 ft) wide. Asparagus will rot if it's too wet, so if you have heavy soil, add grit, manure or leafmould, or raise the bed. If it's acid, add lime the following spring to get pH7.

Sowing and planting

Seed takes three years, so it saves time to buy one- to three-year-old crowns. The one-year-olds are said to establish better in the long run. In spring, build ridges in the prepared trenches and place the crowns in two staggered rows 45 cm (18 in) apart. Straddle them across the ridges with the roots going downwards. When you fill the trench, the crowns should be about 15 cm (6 in) below ground.

Cultivation

Hand weed regularly. Don't cut the asparagus in the first year. In the autumn, cut the leaves down to the ground when they turn yellow. Give the crop a good thick mulch of well-rotted manure or compost. Feed with some fertilizer the following spring – seaweed would be a good choice. Watch out for late frosts.

Harvesting

In their third year, your patience will be rewarded. The asparagus can be harvested for six weeks from May to June. Use a sharp knife to cut the spears just below the ground, taking care not to damage the crown. From mid-June onwards, allow the crop to recover until the following year.

Problems

Usually trouble free, though slugs and snails can be a nuisance, as well as asparagus beetle and violet root rot. Late frosts can spoil an early crop if you don't give it protection.

Recommended varieties

'Backlim' F1 (AGM) – Large spears, high yield. An all-male hybrid.

'Connover's Colossal' (AGM) – A favourite since the 19th century. Early, large spears with a fine flavour. Best in light soils.

'Giant Mammouth' – Similar to 'Connover's Colossal' but tolerates heavy soil.

'Gijnlim' F1 (AGM) – An all-male hybrid. Consistent, prolific and early. Purple tips.

**'Jersey Giant'* – An all-male hybrid producing fat, tender spears.

'Lucullus' F1 (AGM) – Late but high yielding. Elegant, slim, medium-sized spears.

Crown to harvest: one or two years. *Seed to harvest:* three to four years.

CHAPTER 2

GROWING FOR SHOWING

FLOWER SHOWS

Grand flower shows were very much a feature of the 19th century as the Victorians had a huge curiosity for the natural world. The Royal Horticultural Society Great Spring Show, the forerunner of the Chelsea Flower Show, was launched in 1862. The heyday of the local flower show, however, came much later. Digging for Victory during World War II proved an inspiration for many, some of whom had taken up gardening for the first time, and they wanted an outlet for their new-found talents.

Most towns and villages still hold an annual show. The name 'flower show' is something of a misnomer as they are usually more about vegetables. They often include crafts, art, jam and cake making as well. Some have classes for children – for a garden in a seed tray or an animal made out of vegetables. Sometimes there are also classes for beginners – for the tallest sunflower or the biggest marrow, traditionally depicted arriving in a wheelbarrow.

Many of the larger allotments, or groups of allotments, have their own shows and the revival of interest in allotments is reflected in the showing scene. In recent years there has been a positive drive from many quarters to raise the allotment profile by putting on a show. 'Allotment Week', set up in 2003, encourages sites across the country to have an open day or a show during August.

Locals

The majority of shows are run by small horticultural or flower societies. Among the societies, some 3,000 are affiliated to the Royal Horticultural Society, so competitors can pick up a prestigious RHS medal. The RHS *Horticultural Show Handbook* is the show bible. It lays down the criteria and standards that are closely followed across the country. It offers guidance to both exhibitors and judges on every aspect of showing flowers, fruit and vegetables.

Nationals

Competition gets fiercer at the 'nationals'. As a member of a national society, you are invited to conferences, receive a society journal and can compete with the best

POT LEEKS

Pot leek competitions go back to the 1880s in the north east of England. At that time every self-respecting pub and club in the mining villages would run a competition. The Pot Leek Society has an annual national show (usually in Sunderland in September), an annual general meeting in March, and winter 'talk-ins' where the members are encouraged to cross-question the champions. Growing short, hefty pot leeks is a highly specialized art and unlike any other branch of showing. Points are awarded for the cubic capacity of the blanched shaft, from the basal plate to the 'tight button', the point at which the lowest leaf forks.

growers in the country. The rules closely mirror those of the RHS but there are some minor differences in the pointing system. Whereas in most flower shows it is the honour that counts, at national level the prizes are substantial. As a member you join an élite.

The National Vegetable Society (NVS), formed in 1960, has five affiliated groups around the country in the North, the South, the Midlands, Scotland and Wales. Each has its own show once a year and will host the prestigious National Show every fifth year. The NVS train their judges, set them exams and supply them for local shows.

SHOW SUCCESS

Showing is both an art and a science. The aim is to achieve the peak of perfection for a particular day despite the vagaries of the weather, to get your exhibits to the show undamaged and spanking fresh, and to present them in the specified manner. Studying the rules and following them to the letter is vital to success.

It is important to note that where the number of vegetables is specified in the particular schedule, it is strictly adhered to. Neither more nor less may be shown. If you get it wrong, you could be disqualified as NAS (not according to schedule).

The points system

The pointing system takes account of the skill involved in cultivating different types of vegetables. The tricky cauliflower can achieve 20 points, the comparatively easy cabbage 15, and a Chinese artichoke can get no more than ten points. This is highly relevant to your odds if you are doing a mixed exhibit or 'a collection', considered to be the most prestigious of the classes. The RHS recommends entering high quality vegetables that can't get the maximum score rather than poor ones that could.

Tricks of the trade

Many believe that there is a lot of trickery involved in showing. In truth, success comes from close attention to detail, patience, experience and horticultural skill. Most of the judges have been exhibitors themselves and there is no question of pulling the wool over their eyes. There are a few legitimate tricks of the trade, however, and some of these are mentioned below.

Choosing stock

When starting out, most people choose seed recommended for showing from seed merchants with household names. As competition gets hotter, they may buy from the specialist exhibition seed companies. These suppliers concentrate on following the shows and field trials to pick out strains with winning potential. Exhibitors also buy seed or cuttings from the champions' own stock.

GROWING GIANTS

A common misconception is that size is more important than quality in showing, but this is not generally true. The exception to this is the Giant Vegetable Championships. These are quite separate from the rest of the show world and they don't follow the RHS or NVS rules. The

Specialist seed suppliers, interesting small advertisements, lists of all the shows and other useful information can be found in ***Garden News***, the showman's newspaper, available over the counter.

annual National Championships which have been held at the Bath and West Showground, Shepton Mallet, since the 1980s are a public entertainment, drawing in huge crowds and big money.

Described as "the battle of the giants", exhibits are truly gargantuan. Lifting equipment is needed to move the pumpkins and four strong men, each carrying a corner of a blanket, carry them into the tents. The average length of a parsnip at the show is in the region of 4.5 m (15 ft), a carrot is 3.6 m (12 ft), and the weight of a standard marrow is 50 kg (100 lb). Vegetables can only be grown to this size with the luxury of a glasshouse and with more attention and specialist skill lavished upon them than most of us could provide. Competitors are drawn by the big prizes, the chance to be in the Guinness Book of Records and the challenge.

In the normal run of shows, too large can be considered defective. Under RHS rules there is, for example, a class for "Carrots, long" with a specified minimum length of 30 cm (12 in), and "Carrots, other than long". A potato that is too large loses marks.

VEGETABLE VARIETIES

Some vegetables are more popular for showing than others. Here are some popular varieties, with hints and tips for success.

Showing potatoes

An example of the RHS schedule for potatoes is:

POTATOES

Meritorious

Medium-sized of generally between 170 g and 225 g per tuber; shapely, clean, clear-skinned tubers. Eyes few and shallow.

Defective

Tubers that are very small or very large or mis-shapen or have damaged, speckled or patchy skin or have many or deep eyes.

Condition:	*5 points*
Size:	*3 points*
Shape:	*4 points*
Eyes:	*3 points*
Uniformity:	*5 points.*
TOTAL:	*20 points*

Six for a collection, five for a single dish

As you will see the most points are for quality (5) and uniformity (5) and you are penalized for very large tubers. If you are aiming for a mixed exhibit, potatoes are a good way to clock up the points. In many shows there are two classes for potatoes (kidney and round) and in some shows there are four different classes (kidney, white and coloured; and round, white and coloured) giving you the possibility of 80 points straight off.

GROWING SHOW POTATOES

The soil should be carefully prepared the autumn before the show. Choose a potato variety that suits your soil, and aim to have sturdy, sprouted, egg-sized tubers ready to plant out in March. Some exhibitors rub off any weak shoots, leaving about three or four to develop.

Some people plant the tubers out in trenches lined with well-rotted manure or compost. They cover them with varying mixtures of sifted peat, leafmould, old manure, grass clippings and old potting soil. Others avoid soil altogether and grow them in sedge peat or mushroom compost.

Some growers use stakes and string to keep the foliage upright as it grows. Some spray with Bordeaux mixture against blight if the summer is damp. All weak, spindly shoots are removed as they appear and the ridges are kept well hoed to prevent greening.

HOW AND WHEN TO HARVEST

It is advisable not to leave the tubers too long in the ground. If you dig up a perfect specimen for the kitchen before time, you can bury it again, marking the spot. Usually the whole crop is lifted in August or a week before the show. Dig up with extreme care. Cut off the leaves and store the tubers loosely in a bucket filled with the growing medium. Keep in the dark with some air circulation.

As near to the last minute as possible, soak them in water and rub the mud off gently with a sponge. Clean the 'eyes' with a cotton bud. Rinse and pat dry gently. Lay them out in a row in gradations of size and make the final choice to fit the criteria in the schedule. Uniformity is one of the most difficult of challenges and it is as well to grow plenty to get a matching set of a few. Wrap them up individually in tissue paper and newspaper.

Take a black cloth to the show, and paper plates to arrange your produce on if they are not provided. Arrange the tubers in a circle with eyes outwards. Cover them with paper to keep out the light until the very last minute.

Showing cabbages

An example of the RHS schedule for cabbages is:

CABBAGES

Meritorious

Shapely, fresh and solid hearts with the surrounding leaves perfect and bloom intact and of good colour. Size and shape according to cultivar with approximately 75 mm of stalk.

Defective

Hearts that are soft, split or lack freshness, or are pest-damaged.

Condition including solidity:	*5 points*
Size and shape:	*4 points*
Colour:	*3 points*
Uniformity:	*3 points*
TOTAL:	*15 points*

Three for a collection, two for a single dish

GROWING SHOW CABBAGES

Choose your varieties carefully and take great care to avoid pest damage. Timing it right for the exhibition can be tricky so hedge your bets by making several sowings a week apart. Some exhibitors retard their cabbages that are ready too soon by pushing a fork under them and levering them up slightly. Others dig them up, hang them upside down in a cool shed and give them a fine spray of water daily.

HOW AND WHEN TO HARVEST

If you've got the timing right, lift cabbages the night before the show. Wash the roots with a strong jet of water and rinse the leaves carefully. Stand the cabbages in buckets with the roots in water overnight. On the day of the show, the roots can either be left on and wrapped in a damp cloth until the last minute, or cut back depending on the particulars of the schedule. You can strip off a few of the outer leaves if necessary.

Showing runner beans

An example of the RHS schedule for runner beans is:

BEANS, RUNNER

Meritorious

Long , slender, straight, fresh pods of good colour with stalks and no outward sign of seeds.

Defective

Pods that are short, broad, mis-shapen, damaged, tough or stringy, of poor shape and colour or that have prominent seeds.

Condition:	*5 points*
Size and shape:	*5 points*
Colour:	*4 points*
Uniformity:	*4 points*
TOTAL:	*18 points*

Twelve for a collection, nine for a single dish

GROWING SHOW RUNNER BEANS

Runner beans are a popular entry, so competition can be steep. The soil should be carefully prepared the autumn before to allow for a good root run to a depth of 60 cm (24 in) and have plenty of organic matter incorporated. You need to sacrifice quite a lot of the crop to produce beans for exhibition. Allow them extra space – at least 45 cm (18 in) between plants so they can luxuriate. When the bines reach the tops of their supports, stop them by

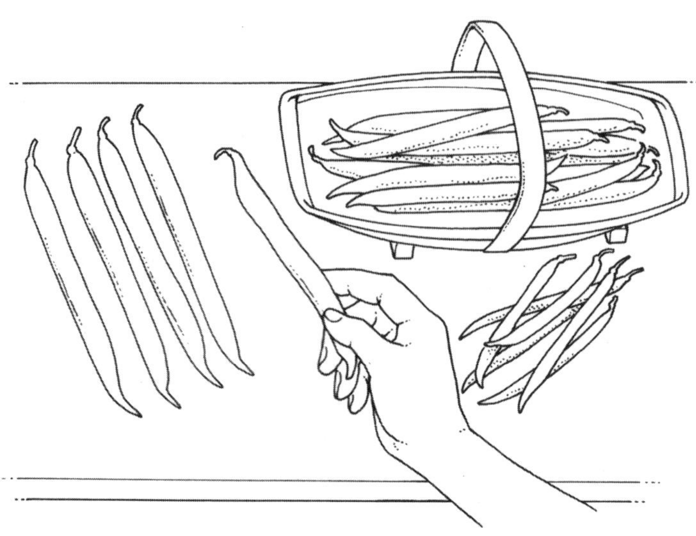

Comparing beans for uniformity

pinching out the top growth just above the last flower truss and stop all laterals at two leaves above a flower truss.

When the beans are about 5 cm (2 in) long, thin the trusses to the three most promising pods. Remove any others that don't come up to scratch. To get the largest possible specimens you may want to thin again later. Two weeks before the show, choose the best and straightest beans and mark them by tying a bit of raffia around them. Some exhibitors run the beans between finger and thumb to 'stretch' them when the seeds are forming. Keep well mulched, be attentive to watering and feed with a high potash fertilizer.

HOW AND WHEN TO HARVEST

To achieve uniformity, the larger beans can be harvested a few days before the show giving the smaller ones time to catch up. They should be picked with some stem and kept in a cool dark place with the stems in a little water. Change the water daily and every two or three days trim the stems by a fraction.

Some exhibitors straighten them further two days before the show by laying them out on a flat surface, covering them with brown paper under a damp towel and placing a board on top.

Lay your collection of beans out to choose the most uniform. Wrap them in a damp cloth on the way to the show and keep them covered with it until the last minute. They are usually laid out side by side with the stalks pointing inwards.

Showing beetroot

An example of the RHS schedule for beetroot is:

BEETROOT, GLOBE

Meritorious

Spherical, of approximately 50–70 mm in diameter with a tap root, smooth skin and flesh of a uniform dark colour.

Defective

Specimens that are mis-shapen, tough, have too many roots or rough skin, or flesh that has rings.

Condition:	*5 points*
Size and shape:	*4 points*
Colour of flesh:	*3 points*
Uniformity:	*3 points*
TOTAL:	*15 points*

Six for a collection, six for a single dish

GROWING SHOW BEETROOT

Exhibition beetroot can be given a good start by using John Innes number 2 potting compost plus some organic matter. It must be very well rotted as, if it is at all fresh, the beets may fork. Opinions vary on the mix, as they do most things when it comes to exhibiting. Some swear by a dressing of agricultural salt, as beets are a seaside plant. Sow the seed in April or May and grow on the beets in the normal way. Watch out for beets pushing their shoulders out of the ground. If this happens cover them with soil to prevent them getting dry.

HOW AND WHEN TO HARVEST

Measure the shoulders before harvesting. When digging them out take great care not to damage the tap root or to bruise or damage the flesh as beetroot bleeds. Wipe the roots carefully with a damp rag and trim off any little rootlets on the sides without touching the taproot. Some exhibitors believe that soaking them in salty water for 15 minutes deepens the colour. Trim off any tired outer leaves. Rub off any corkiness from the shoulders with a saucepan scrubber. Usually the leaves are cut back to around 7.5 cm (3 in) but check the schedule. Transport them to the show in a damp cloth. At the show, cut off the tops to the specified length and tie with raffia or string. They are generally exhibited on a plate with the roots facing inwards and given a little spray of water to freshen them up before judging. The judges will cut one in half, so it is well to check for rings and good colour before submitting an entry.

Showing carrots

An example of the RHS schedule for carrots is:

CARROTS

Meritorious

Fresh, long roots of good shape and colour. Size to exceed 300 mm from the crown to a point where the root measures 50 mm in diameter; skins clean and bright.

Defective

Roots that are coarse, mis-shapen or with coloured crowns; fangy, dull, blemished or poorly coloured.

Condition:	*6 points*
Colour:	*5 points*
Size and shape:	*4 points*
Uniformity:	*5 points*
TOTAL:	*20 points*

Three for a collection, three for a single dish

GROWING SHOW CARROTS

To achieve long exhibition carrots you need to use seed of giant strains and provide perfect growing conditions. This is either achieved with the crowbar method of digging a hole some 1.35 m (4 ft 6 in) deep and then plunging a crowbar into it and turning it to make an inverted cornet shape. This can be filled with light and sandy soil that carrots like, with some balanced fertilizer mixed in. Many growers use the seaweed fertilizer Seagold and bonemeal.

For the late summer and early autumn shows, station sow seed in April and thin to one plant. Cover with cloches if the weather turns cold in the first few weeks. It is important to make sure that the carrots never dry out as the roots may split. Regular moderate watering is the key. Draw soil over the crowns of the carrots to avoid greening or corkiness.

Some experts say that the best colour comes from keeping them at a temperature of around 19°C (66°F) in the day and 7°C (46°F) at night. A little shade therefore is desirable in the heat of summer and cover with cloches at each end of the season at night.

HOW AND WHEN TO HARVEST

To save time, mark out your shortlist of carrot candidates the week before the show. Gently ease away the growing medium from around the crowns and mark the most hopefuls with plant labels. Cover them afterwards to prevent 'greening'. Two days before the show, saturate the soil by plunging a hose down by 1 m (3 ft 4 in). The following day the carrots should pull out with ease.

Hose down the carrots and wash off any remaining mud with a sponge, taking care with the crowns. Remove any fine root hairs that have grown along the length of the carrot but leave the tap root intact. Wrap the carrots in damp cloth. Lay them in a box to transport to the show.The NVS recommends placing small pieces of sponge between each layer to prevent any movement that might cause damage. Usually the foliage is cut off 2.5 cm (1 in) above the crown, but check the schedule.

Showing cauliflower

An example of the RHS schedule for cauliflower is:

CAULIFLOWER, INCLUDING WHITE HEADING BROCCOLI

Meritorious

Heads with symmetrical, close, solid white curd, free from blemish, stain or riciness, with approximately 75 mm of stalk.

Defective

Heads of irregular shape, beginning to open, yellow, blemished or stained or showing leaf in the curd.

Condition including solidity:	*5 points*
Colour:	*5 points*
Size and shape:	*5 points*
Uniformity:	*5 points*
TOTAL:	*20 points*

Three for a collection, two for a single dish

GROWING CAULIFLOWER FOR SHOWING

It is difficult to time for perfection to the day and the judges look for freshness, so sow a few seeds several days apart. Only late summer or early autumn cauliflowers will be ready at the time of the shows.You will need to grow quite a few to come up with uniform heads.

HOW AND WHEN TO HARVEST

Harvest as near to the show as possible, on the day of the show if you can. Cut with about 15 cm (6 in) of stalk. Keep the cauliflowers covered until the last minute. Cover the curds with tissue paper and put each head into a polythene bag. Check the schedule to see how much stalk to cut off. Usually all but the inner leaves are cut off. Cut the outer leaves at the last minute.

Showing leeks

An example of the RHS schedule for leeks is:

LEEKS, BLANCHED AND INTERMEDIATE

Meritorious

Clean, firm, solid, parallel-sided, long barrels with no sign of softness or splits, with a tight button and free from bulbing and ribbiness.

Defective

Leeks that are soft, thin, tapering, short-shafted, imperfectly blanched, discoloured or bulbous, or that have diseased or damaged leaves.

Condition:	*6 points*
Size and shape:	*6 points*
Colour (blanch):	*3 points*
Uniformity:	*5 points.*
TOTAL:	*220 points*

Four for a collection, three for a single dish

GROWING LEEKS FOR SHOWING

Many exhibitors grow their own strains and propagate them from side shoots or suckers.The other method is to take 'pods', or miniature plants, that appear amongst the ripened seeds. They are removed in early winter and potted up in the greenhouse.These can also be bought – *Garden News* advertises them as 'blanched pips'.

They are often grown on within a wrapping of cardboard covered in polythene or sections of roofing felt to get a good length of blanch. Some exhibitors grow them in peat rather than soil as it is clean and spongy. They are carefully uncollared from their wrappings to be measured from the root base to the 'tight button', the point at which the lowest leaf divides. This is because some schedules specify that the blanch of intermediate leeks should be over 15 cm (6 in) and under 35 cm (14 in) in length.

HOW AND WHEN TO HARVEST

Lifting leeks with the roots intact and without damage is something of a challenge. Tie the tops so that the weight of the leaves won't cause them to tear. Scrape the soil away, then cut down vertically into the soil with a knife to release the leeks. Wash off as much soil as you can with the hose. Take them into the kitchen and lay them on the draining board on a folded towel with the roots under a running tap. The little root hairs (beard) should be combed straight.

You can take the outer leaves off if you wish. The RHS advises to avoid excessive stripping as it exposes ribbing. Stand them with their roots in water and keep in a dark place until it's time to get ready for the show.

The NVS recommends wrapping the leek barrels and roots in a damp towel to keep them cool and to stop the roots drying out. A layer of bubble wrap goes over the towel and is topped by a layer of black plastic dampproof course to keep them rigid. Wrapped like this they will keep fresh for a couple of days on the garage floor if necessary and will travel without danger of damage in the boot of a car. Leeks are usually lined up flat on the show bench. They look good on a black cloth.

Showing peas

An example of the RHS schedule for peas is:

PEAS

Meritorious

Large, fresh pods of good colour with bloom intact, free from disease or pest damage and well filled with tender seeds.

Defective

Pods that are small, not fresh or of poor colour or having very imperfect bloom, or which are diseased or pest damaged or poorly filled or containing seeds that are old or maggoty.

Condition:	*7 points*
Fullness and size of pod:	*5 points*
Colour:	*3 points*
Uniformity:	*5 points*
TOTAL:	*20 points*

Twelve for a collection, nine for a single dish

GROWING PEAS FOR SHOWING

To get twelve matching, perfect pea pods you will need to sacrifice quite a lot of the crop. Weaker pods and any that are ripening too early need to be removed from the plants regularly. In hot weather you may need to shade the bines. Start selecting the best pods a couple of days before the show.

HOW AND WHEN TO HARVEST

The peas should be harvested at the last possible moment with the aid of sharp scissors, leaving about 7.5 cm (2 in) of stalk.

Handle by the stalk to avoid removing the bloom from the pods. Hold them up to a strong light to check that there are no maggots hidden in the pods and to count the peas. The stalks are cut down further on arrival. They are usually exhibited in a wheel formation with the stalks pointing inwards.

PREPARING FOR THE SHOW

As most of the preparation can't be done until the last minute, it is a good idea to do what you can to get ahead. One of the best ways to prepare is to go to a few shows beforehand to get some ideas on presentation and familiarize yourself with the way they are run.

Presentation

Vegetables are usually displayed on black cloths. Velvet, if allowed, is generally agreed to show the vegetables off to best advantage.

A variety of staging props can be used under the black cloths to present the vegetables in the right position. There may be a backboard to show off vegetables that are best displayed upright. Sometimes there is a sloping board to provide a better perspective on smaller vegetables like potatoes and tomatoes.

Displaying tomatoes on a cone

A spiked board is useful to secure vegetables that are displayed standing up straight. Another method is to tie them to hangers with a little bracket to take the weight.

Small blocks of wood of varying heights from 2.5 cm (1 in) and up are sometimes used to display the different dishes. Some exhibitors use florists' baskets decorated with parsley, or attach pod vegetables such as peas to wire cones with bouquet pins.

The show kit

Write out your labels and start collecting together boxes and packing materials early. Exhibitors need a tool box containing hammer and nails, florists' wire, wire cutters, secateurs, a penknife, string, a ruler, a water sprayer, a notebook, pencil and pens amongst other essentials.

Tactics

Aim to arrive early as most shows work on a first come first served basis. The NVS recommends choosing a space at either end of the show bench if you can. In that position, you can place your weakest exhibits on the outside and avoid being judged directly against the superior set that might turn up next to yours.

Displaying turnips in a 'Welsh Hat' basket

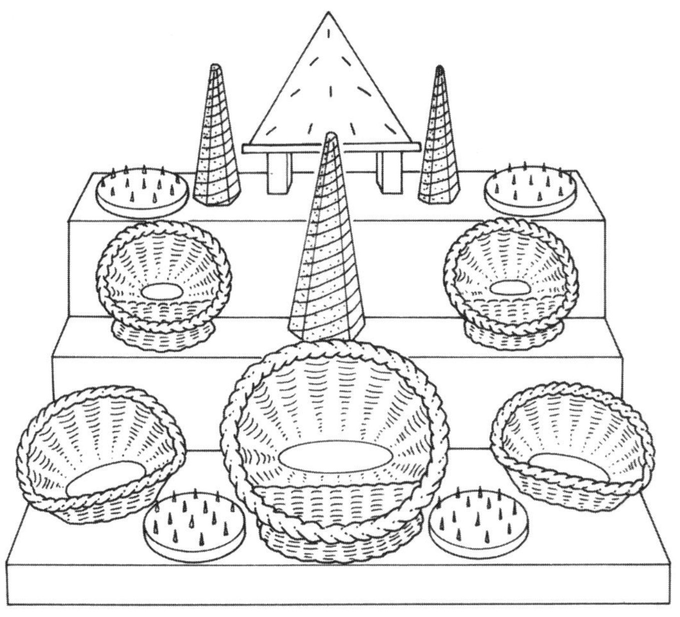

A selection of cones, baskets and spiked boards for displaying vegetables

ALLOTMENT PLOT COMPETITIONS

There are many in-house competitions for the best kept plot and most councils run one between different allotment sites and individual plots. *The National Society of Allotment and Leisure Gardeners* (NSALG) and *Garden News* run a national allotment plot competition.

The RHS *Horticultural Show Handbook* offers guidelines to judges on condition of plot, good workmanship, quality of crops, originality, ingenuity in overcoming local problems, visual aspect and the condition of garden sheds. They wisely point out that as allotments allow enthusiasts for a particular plant to indulge their passion, so they should be judged according to the standard and quality of cultivation.

One of the great strengths of allotments is their individuality, so unlike the flower shows, there can be no strict pointing system. NAS (not according to schedule) not does not apply.

CHAPTER 3

GARDENING TECHNIQUES

WEEDS

Weeds are great survivors. They breed prolifically and have the knack of finding the particular conditions they need to carry on doing so. It is unrealistic to think you can get rid of them entirely, especially on an allotment where seed can be blowing in from all sides. In fact it's a good idea to keep a few of those that bring in bees and butterflies in odd corners or have other uses. Nettles, for example can be used to make good liquid manure.

The aim is control rather than elimination. You don't want them, good or bad, where you are going to sow seed or amongst your crops. They will compete for nutrients, water, light and space. The first step is to identify them by type. They divide into annuals and perennials.

Annual weeds

Annual weeds grow, flower, seed and die in the course of one season. They propagate themselves by producing quantities of seed, sometimes in their thousands. To get on top of them it is vital to prevent them from flowering and setting seed. If you are really stuck for time, be sure to dead head. As they generally live in the top few centimetres of the soil, annual weeds can be kept down by hoeing and hand-pulling. Through the growing season, keep your hoe to hand so you can catch them young, before they make much root. If you feel you are losing the game, give your plants a head start against the competition by transplanting, using pre-germinated or chitted seed, or by fluid sowing.

Weed seeds lurk dormant in the soil. When you disturb them and bring them up near the light they will germinate.

THE STALE SEEDBED

One way to get ahead of annual weeds is to make a stale seedbed. Prepare the bed for planting and let the resulting weeds germinate. Hoe them off before sowing your seed and with luck your seeds should be up before any more weeds appear. If the weather is cold, warm the soil by covering it with polythene or fleece for a couple of weeks to encourage the weeds to show themselves first.

Mulching is a godsend. It really does help by excluding light and smothering them. A 5-cm (2-in) depth of organic mulch between plants will dispose of most annual weeds. As the mulch is of loose texture any rogue seedlings can be pulled out with ease. A good leaf cover of crops will shade them out, too. Aim to keep the ground covered one way or another.

Perennial weeds

Perennial weeds go on from year to year and present more of a challenge. On land the size of an allotment you need to take sweeping measures if you have a serious problem with them. Most can spread from their roots as well as from seed. Dandelion and dock grow massive taproots. Couch grass, ground elder, horsetail and willow herb make a tangled underground network while the roots of coltsfoot and bindweed are deep and spreading. All are difficult to dig out cleanly and will sprout from the tiniest section of root left in the ground. Rotavating them will only make things worse as it chops the roots up along with the gardener's best friend – the worms. Another disadvantage is that it can damage the soil structure. If you don't plan to use all the land at once, consider digging and hand-weeding some areas for immediate use and smothering the weeds in others. Deal with them step by step, even though it may take several years to eliminate them completely.

Perennial weeds on a large scale

If your plot is covered in grass and weeds, scythe or strim it down and keep it short. Only small rosette weeds, like daisies, will survive constant mowing. When you are ready to plant, take off the turf, bury it upside down about a spit down and cover with the topsoil. It will rot down into good loam. Another method is to stack turfs upside-down to make a turf stack.

Excluding light No plant can live without light. If you have tall weeds, scythe or strim them down. Cover the area with heavy black plastic buried at the edges and weighed down with stones or bricks. The length of time you need to keep the cover on depends on the particular weed. Some will give up the ghost after a year while the most persistent can take three. If you make slits to allow in water or use porous horticultural plastic, the land needn't be wasted. You can plant through the cover by making cross slits. It is best to grow plants that are a fair size and quite vigorous by nature.

Hessian-backed carpet is widely used for excluding light and works well, though not all sites permit it. Avoid the foam-backed type as it will eventually break down and

A mulch of heavy black plastic will kill bad perennial weeds by excluding light. Vigorous plants can still be grown through slits.

be difficult and unpleasant to remove. Other possibilities are large sheets of heavy cardboard of the type used by removal, kitchen or bathroom appliance firms.

Flame gunning This has been used in agriculture since the 19th century. A flame gun is not designed to burn the weeds but gives off the right heat for the cell structure to change and rupture the cell walls several hours later. You can tell if the treatment has been effective by pressing a leaf between finger and thumb. If it has worked it will leave a dark green fingerprint. Weeds are flame gunned at about 71°C (160°F) for a single second.

In agriculture, flame gunning is generally used for clearing small weeds in the carefully calculated time between sowing a crop and the seeds emerging. It has the advantage of not disturbing the soil which will only bring more weed seed up to the light to germinate. Tough perennials may need several treatments. Obviously there are always dangers attached to fire and you may need permission to use a flame gun from the site manager.

THE SOIL

Good soil is bursting with microscopic life – fungi, algae, bacteria, worms, vegetable and animal remains, air and water. It should be a life-giving cocktail. Look after your soil and it will look after the plants.

If you have been dealing with lush stinging nettles, chickweed and dock you can be fairly sure that your soil is fertile. A thriving worm population is an excellent sign and will tell you that the soil is friable and will be easy to work. Ideally there should be a two or three worms on every spadeful you dig. Healthy soil has a pleasant, earthy smell.

Feel the texture

You can learn quite a bit about your soil by picking up a small handful half an hour after it has rained. Sandy soil will feel gritty, silt will be silky, and clay will feel sticky. Loam – the gardener's dream soil – contains roughly equal amounts of all three.

If the handful of soil will mould into a ball, it will be silt or clay. Clay goes shiny when rubbed. Chalk will slip through your fingers, while peat is dark and crumbly.

Clay soil Clay soil is rich in minerals and nutrients. A heavy soil, the disadvantages are that it can become waterlogged, is sluggish to warm up in spring, tough going to dig and cakes in the heat. The brassicas, which like firm planting, do well on clay.

Treading on clay will compact it and make it airless, so consider making raised beds which you can reach across. Dig it over in the autumn and let winter weather get at it to help to break it down. Piling in sharp sand, grit and organic matter will help incorporate air and some free-flowing drainage through it, and transform it into a workable, good soil.

Whatever type of soil you have, the addition of generous quantities of organic matter will improve soil fertility, drainage, water retention and texture right across the board.

Sandy soil Sandy soil is free-draining and easy to work. It warms up quickly in spring. It's a light soil which is particularly good for salad crops, most roots and legumes. However, it can be so free-draining that nutrients are washed away, or leached, as through a sieve, by rain. Sandy soils are often low in potash.

Adding well-rotted compost or manure will help to bind the particles together and retain nutrients and moisture. Leave any digging until spring and keep it covered through winter to minimize leaching

Silt Silt is an alluvial soil, typically from riverbanks, and lies somewhere between clay and sand. It is easily compacted but holds on quite well to nutrients.

Chalk Chalk is a free-draining, poor soil. It's full of lime, which makes it alkaline and inhospitable to many plants. Organic matter will help to bulk it up and counteract the pH. It is often low in potash.

Peat Peat, which comes from wetlands, is light and easy to work, fertile but acid. It retains water when wet but dries fast when it has a tendency to blow away. Adding lime will make it more alkaline and organic matter will give it substance and weight.

Test the pH

A simple pH test will tell you whether the soil is neutral, acid or alkaline. The micro-organisms which provide vital nutrients for the plants do not prosper in either extreme. The optimum for most plants is pH5.5–7.5.

Collect three or four small samples of soil from different parts of the plot. You can send them off for a full analysis (as advertised in gardening magazines), but a cheap kit from a gardening centre is probably all you need. A pH of 7 is neutral, a pH of above 7 is alkaline, and below 7 is acid.

If the soil is too acid, add lime to counteract it while adding calcium to the soil. Ground limestone (calcium carbonate) and dolomitic limestone (calcium magnesium carbonate) are the organic choices. Do not add it at the same time as manure as it will react against it. The general practice is to lime in autumn and manure in spring. If the soil is too alkaline, garden manure and compost will send it in the right direction. While you can tip the balance and improve the soil, you cannot completely change its character. However, you can choose plants that prefer your particular soil type.

The soil profile

If you are being thorough, or have worries about drainage, it is worth digging a small hole about 1 m (1 yd) deep. This will reveal some interesting horizontal bands – the topsoil, the subsoil, broken rock and the bedrock below.

Topsoil The first layer is noticeably darker than the rest. It is the layer that feeds the plants. Most vegetables will be happy with a layer of topsoil 38 cm (15 in) deep.

Have a look at the soil structure. This is the way particles clump together to form crumbs. If you can see plenty of holes and cracks in the exposed face you can be sure that there is plenty of air going through. If not, then you need to open it up by adding organic matter or raising the beds.

Subsoil The underlying subsoil is lighter in colour. It contains few plant nutrients but its structure affects drainage. It is important that water can flow away and air can get to the roots of plants.

Pour some water down the hole to see

if it runs away. If it doesn't it could be due to compacted airless topsoil or an impermeable barrier in the subsoil known as a hard pan. This can usually be broken up with a pickaxe or loosened with a fork and kept aerated with regular additions or organic matter.

You need a good quantity of compost to get the full effect – some 5 kg (10 lb) of compost per square metre (yard) of soil, applied on an annual basis, should make a big difference.

ORGANIC MATTER

Organic matter is a godsend to gardeners, improving soil structure, adding nutrients and even improving soil pH. It can be incorporated into the soil by digging, or used a mulch on the surface to suppress weeds and retain moisture.

Garden compost

Composting is a speeding up of Nature's own recycling process. As animal and vegetable remains rot down, the population of micro-organisms burgeons. The heap heats up quickly. If it gets hot enough it will kill off weed seeds and pests. As the activity slows and the heap cools, the micro-organisms are joined by beneficial worms and insects.

When the compost is ready to use the volume will have reduced by half, it will be sweet smelling, dark and crumbly and the original contents will be unrecognizable. The end result is humus. It enriches the soil for new cycles and generations of microscopic life. It dramatically improves soil texture, structure, water-holding capacity and drainage.

Compost bins You can have an open compost heap but it won't heat as well and can blow about. The best plan is to have three bins – one ready, one rotting down and one in the making. You need a container without a bottom for the worms to get in and do their work. Small gaps are important for air circulation but you don't want gaping holes as they will let the heat

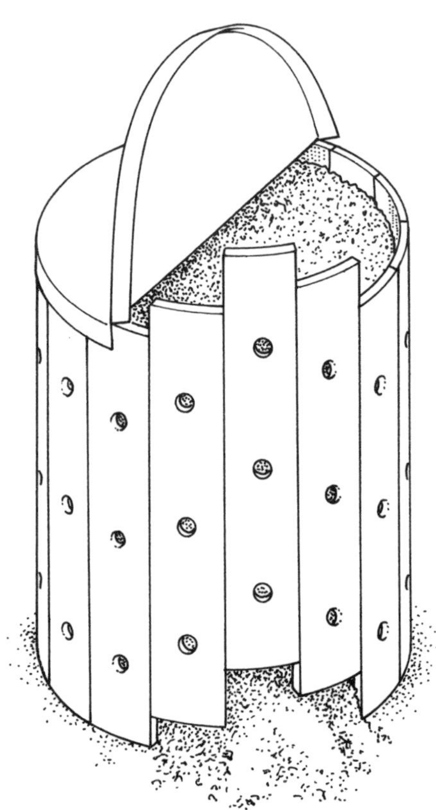

A proprietary tongue-and-groove wooden compost bin

out and dry the compost. You need a lid or cover to keep the rain off and access from the side or top in order to turn the compost and get it out. For fast and effective composting, make a bin of at least a cubic metre (cubic yard).

Wood is the best material for compost bins as it has insulation and 'breathes'. Plastic, including recycled, is commonly used though it doesn't keep the heat as well.

How to compost The bigger the heap and the more you put in at the same time, the faster and more effective you will be.

A HOMEMADE COMPOST BIN

An excellent, no cost, compost bin can be made from three wooden pallets nailed together at the corners with the fourth tied on for the 'door'. Push straw, old sacks or newspaper into the gaps and make a lid from old carpet, heavy plastic or a piece of board.

The faster a compost heap is constructed, the sooner it will become activated. Try picking up extra material from outdoor markets and greengrocers. Or why not club together with other allotment people to make a joint heap.

Start off the heap with something coarse and twiggy to let in air. Collect and save a good assortment of materials next, aiming for about half green materials to half dry. Chop up woody materials into short lengths and crush the tough stalks. Cut fabrics up into small pieces. It's worth filling a few dustbins full of different things before you start.

Water dry materials – straw, paper, and cloth – and squeeze them out. If you have too much green (particularly mowings) you will end up with silage. You want moisture but not sogginess as it will make the heap putrefy.

WHAT TO COMPOST

In theory, you can compost anything organic – kitchen scraps, tea and coffee, eggshells, wood ash, hair, newspaper and cardboard, natural fabrics, garden prunings and weeds.

If you are not sure that you are going to get maximum heat, it is prudent to leave out perennial weed roots, weed seeds, fish and meat (which might attract rodents) and diseased material. The top growth of potatoes often contains potato blight and potatoes may sprout again so don't take the risk. Brassica roots may have clubroot. Cat and dog faeces should never be used for compost where you are growing vegetables or where there are children about. I would also leave out evergreen prunings which are slow to decompose and often poisonous.

To get full value from your compost when it is ready, lay it on the soil in early spring.

Mix your ingredients together, shovel it in and let it settle by itself.

A couple of turnings at intervals of a few weeks will make things go even faster and give you the chance to check on progress. Dig out the whole heap onto a plastic sheet and add whatever it seems to lack – more water or green material if it's too dry, more shredded dampened paper, rags or straw if it's too wet. Overall it should be about as damp as a squeezed sponge. Mix it up again, add some more activator and fork it back.

The composting process takes a couple of months in summer and quite a few through winter.

Trenching compost If you don't have enough material to make much difference to the heap, bury kitchen and garden waste about 30 cm (12 in) deep in a trench, cover with soil and allow it to rot down for a few weeks. Make holes, fill them with compost and plant greedy feeders like courgettes through it. Many people put a layer of mowings or a few comfrey leaves in potato trenches before planting.

Sheet composting This looks unsightly but is another way to make use of small amounts of kitchen and garden waste if you haven't got a compost heap on the go. Lay thin layers on the soil surface between vegetable rows and let it rot down where it lies.

COMPOST ACTIVATORS

A compost activator, added every 15 cm (6 in) or so, will speed things up in a compost heap. A little rotted compost from an old heap works brilliantly. Farm manure, human urine, nettles, seaweed meal, poultry or pigeon droppings, comfrey leaves and blood, fish and bone are excellent too. Lime helps if the heap smells sour or if you are on acid soil.

Farmyard manure

It is worth every effort to get a good supply of farmyard manure.

Never use fresh manure as it will scorch the plants, rob them of nitrogen and may be full of pathogens, even chemicals or hormones. Once rotted down, however, it will be transformed into a great soil conditioner. Manure on straw bedding is the best. The straw is full of nitrogen from the urine and rots down fairly quickly. It should take three to six months whereas manure on wood shavings will take about a year to rot.

The manure heap The bigger the heap, the better and the more quickly it will rot down. A good-sized heap could be 1.5 m (5 ft) high and wide. You can speed up the process by turning it, sides into the middle. Build it on heavy polythene, flatten it with a spade and cover the top and sides with polythene, carpet or a tarpaulin to create more heat. An extra bonus of covering it entirely is that you will kill off lurking weed seeds.

When ready, it will be dark, crumbly and pleasant smelling. The green light to go is the magical appearance of thousands or bright pink branding, or fishermen's, worms.

Using manure For vegetables, it is best spread on the soil thickly in winter for the frosts to break it down even further. The worms will drag it down into the soil. The one exception is if you need to lime the soil in autumn. It will react with the manure and waste the nitrogen. If this is the case, lime in autumn and spread the manure in spring.

Leafmould

Don't waste the fallen leaves you sweep up in autumn. Leaf mould is another valuable soil conditioner, low in nutrients but great for soil structure. Collect leaves in a cage of chicken wire or in black plastic sacks. Add a little water and make a few holes and then leave it for a season or two. If you shred the leaves by mowing over them first it will hasten the process.

Apart from stamping the pile down as you add more leaves, there is nothing else to do apart from watering it in dry summers. Fungi will break the leaves down slowly.

Parks departments are a great source of fallen leaves. Specify that you don't want leaves collected off the roads, as they may contain lead.

In the North, many allotmenteers keep racing pigeons and a few hens. Their droppings are extremely high in nitrogen so they are best used as an activator in the compost heap. Rabbit and guinea pig droppings, which come in small quantities, can be used in the same way.

Other organic composts

- Spent mushroom compost from commercial growers makes a good soil conditioner and mulch for neutral to acid soils. It is a mixture of horse manure, peat and chalk, which makes it alkaline. It's best to leave it in a stack for a few months to get rid of any chemicals or mushroom pests.
- Spent hops are occasionally on offer from breweries. High in nitrogen and potash, they can be dug in without delay if still moist. They also make a good surface mulch.
- Seaweed has an alginate content which binds soil particles together to help the structure. It is rich in trace elements, has much the same balance of nitrogen, potassium and phosphorus as farmyard manure, plus other goodies – vitamins, amino acids, plant hormones and carbohydrates. It boosts plant health, can be used fresh and is an excellent addition to the compost heap. The carbohydrates in it decompose fast, causing a proliferation of helpful micro-organisms.
- Grass mowings are rich in nitrogen and make a moisture-retaining mulch. Use sparingly as they become slimy when they rot down. They can also be used over newspaper to keep down weeds for short periods.

Seaweed makes a great addition to the compost heap and can be collected from the shore or sea if you find it loose, but not if it's growing on the rocks. If you can't get any for free, you can buy proprietary seaweed products.

✦ Fresh bark or wood shavings from factories or the tree departments make luxuriant, weed-free paths.

GREEN MANURES

If you have empty beds that you are not ready to plant, a green manure will make temporary cover and help to keep the area free of weeds. These are fast-growing agricultural crops which are left in the ground for six weeks to a year, depending on the type of plant used. They are cut down when young and dug into the soil, adding fertility and improving soil structure. They are also a good source of composting material.

Green manures are particularly helpful for light soils, as they will prevent erosion and leaching by rain. Some green manures, such as buckwheat and Italian ryegrass, have root systems which will break up heavy ground. The leguminous ones (clovers, winter beans, trefoil and lupins) store nitrogen in their roots. This is released into the soil as they rot down.

Cut down green manures before they flower or when you need the ground. Dig them up, chopping up the foliage with a spade as you go to speed up decomposition. Leave them to wilt for a few days before burying them by single digging. If you are on the no-dig system, leave the residue on the surface to act as a mulch, or compost it. Some of the perennials (clover, trefoil and rye) may grow again and will need to be hoed off or covered with mulch.

Alfalfa (*Medicago sativa*) is a deep-rooting, hardy perennial which grows up to 80 cm (32 in) tall and fixes nitrogen. It is planted

SHORT- AND LONG-TERM GREEN MANURES

Short-term types are fast-growing leafy plants that can be slotted in six to eight week gaps when the ground is cleared between crops. These include fenugreek, mustard, phacelia and buckwheat. The over-wintering green manures – winter tare, grazing rye, winter beans and Italian ryegrass – are sown in early autumn when many vegetables are lifted. They will be dug in the following spring.

Long term green manures – alfalfa, red clover and trefoil – can be left in the ground for a year but should be clipped occasionally to stop them going woody. These are useful for resting overused soil, improving fertility, or simply giving you a break. Keep in mind rotation – winter beans and lupins are legumes, mustard and fodder radish are brassicas.

Phacelia, buckwheat and lupins have flowers which attract beneficial insects. Although green manures are usually dug in before they flower, it might be worth leaving a few in the ground for this reason.

for a full season, either from spring to autumn or from late summer to spring.

Buckwheat (*Fagopyrum esculentum*) is a tender annual, growing up to 90 cm (3 ft) tall. Sow it in April to harvest in June. It has deep roots, smothers weeds, is easy to dig in and copes with poor soil. The flowers draw in helpful hoverflies.

Crimson clover (*Trifolium incarnatum*) and **red clover** (*Trifolium pratense*) are fairly hardy perennials, growing up to 30 cm (12 in) tall. Clover is a nitrogen fixer and loved by bees, so leave a few to flower. Plant in spring to late summer and grow for two or three months, or leave over winter. It is fairly easy to dig in but prefers light soils. Don't use repeatedly as the land may become 'clover sick'.

Allotment people often buy their green manure seed like fenugreek from Asian greengrocers and dried winter (broad) beans from the supermarket. Test them for 'viability' before sowing (see page 150).

Fenugreek (*Trigonella foenum graecum*) is a semi-hardy annual, growing to 60 cm (24 in) tall. It is possibly the best green manure for speed of growth and nitrogen fixing for summer. These are bushy plants with weed-suppressing foliage. Plant in late spring or summer and grow for up to three months in well-drained but moisture-retentive soil.

Lupin (*Lupinus angustifolius*), the agricultural variety of lupin, is sown in spring for a slow-maturing summer crop. It takes two to three months to get to the digging stage. Seeds are sown rather than broadcast. Because these plants are deep rooting, they improve soil texture. They are nitrogen-fixing and effective in suppressing weeds. If left to flower, they are very attractive to beneficial predators. A big disadvantage is that they are poisonous which may well rule them out on the allotment.

Mustard (*Sinapsis alba*) is a tender annual for summer. Rapid and weed-smothering, it needs moisture but isn't fussy about soil. It is said to reduce soil-borne pests and disease. Easy to dig in. A good green manure.

Phacelia (*Phacelia tanacetifolia*) is a semi-hardy annual, growing to 90 cm (3 ft) tall. It has ferny leaves and bright blue flowers attractive to beneficial insects, so leave a few for them. Plant after all danger of frost is past and grow for a couple of months. It will do well on most soils and is easy to dig in.

Ryegrasses Grazing rye (*Secale cereale*) and Italian ryegrass (*Lollium multiflorum*) are hardy annuals used for overwintering and can take almost any soil. They need a couple of months to rot down and are likely to resprout. They are not the easiest

THE BOTANICAL GROUPS FOR CROP ROTATION

PEA AND BEAN FAMILY (LEGUMES)
Papillionaceae

- Asparagus pea
- Broad bean (fava bean)
- French bean
- Lima or butter bean
- Peas, mangetout and sugar snap peas
- Runner bean
- Oriental beans

CABBAGE FAMILY (BRASSICAS)
Brassicaceae

- Brussels sprouts
- Broccoli
- Cabbage
- Calabrese
- Cauliflower
- Chinese cabbage and broccoli
- Kale
- Kohl rabi
- Mibuna greens
- Mizuna greens
- Mustard greens
- Pak choi
- Radish
- Rocket
- Seakale
- Sprouting broccoli
- Swede
- Texcel greens
- Turnip

POTATO FAMILY
Solanaceae

- Aubergine
- Potato
- Tomato
- Okra
- Peppers – sweet, chilli

CARROT FAMILY (ROOTS)
Apiaceae

- Carrot
- Celeriac

to dig up as they have tough fibrous roots. You can save seed for the following year.

Trefoil (*Medicago lupilina*) is a hardy biennial growing to 30 cm (12 in) tall. A summer grower which fixes nitrogen and is one of the few that can cope with some shade and drought. It dislikes heavy acid soils.

Winter beans (*Vicia faba*), or broad beans, are hardy annuals that fix nitrogen. Plant between September and November to overwinter in moist loam soils. They are moderately easy to dig in, but don't forget to leave the nitrogen-rich roots. An excellent green manure that you can harvest and eat. Dry some seed for the following year.

Winter tare (*Vicia sativa*) is a hardy annual growing to 75 cm (30 in) tall. This fast-growing bushy vetch provides plenty of leaf cover. It dislikes drought and prefers alkaline soils. Plant summer or autumn to overwinter. It is reasonably easy to dig in. A five-star green manure for fixing nitrogen and suppressing weeds.

- Celery
- Florence fennel
- Hamburg parsley
- Parsley
- Parsnip

BEET FAMILY
Chenopodiaceae

- Beetroot
- Chard
- Good King Henry
- Perpetual spinach
- Red orache
- Spinach

LETTUCE FAMILY
Asteraceae

- Cardoon
- Chicory
- Endive
- Globe artichoke
- Jerusalem artichoke
- Lettuce
- Salsify
- Scorzonera

ONION FAMILY
Alliaceae

- Garlic
- Leek
- Onions – globe, pickling, Japanese
- Shallots
- Spring onion
- Welsh onion
- Japanese bunching onion
- Tree onion

CUCUMBER FAMILY (CUCURBITS)
Cucurbitaceae

- Courgettes
- Marrows
- Cucumber and gherkins
- Squash
- Pumpkin

CROP ROTATION

The benefits of moving members of the same botanical family to fresh ground each year in rotation was recognized as far back as Ancient Rome and Greece. Rotation controls and prevents a build-up of pests and disease in the soil. These include the real rotters – clubroot in brassicas, eelworms in potatoes and tomatoes, root rots in peas and beans, white rot in onions and canker in parsnips.

The concept is neat and logical. Each vegetable family group has a tendency to the same diseases and attracts the same pests. The most important groups to move round are potatoes, legumes, brassicas and roots. If you have four areas, each year two

It is vital to keep records of what is where each year so you know where you are with your rotation and can treat each bed accordingly, then grow the right crops there.

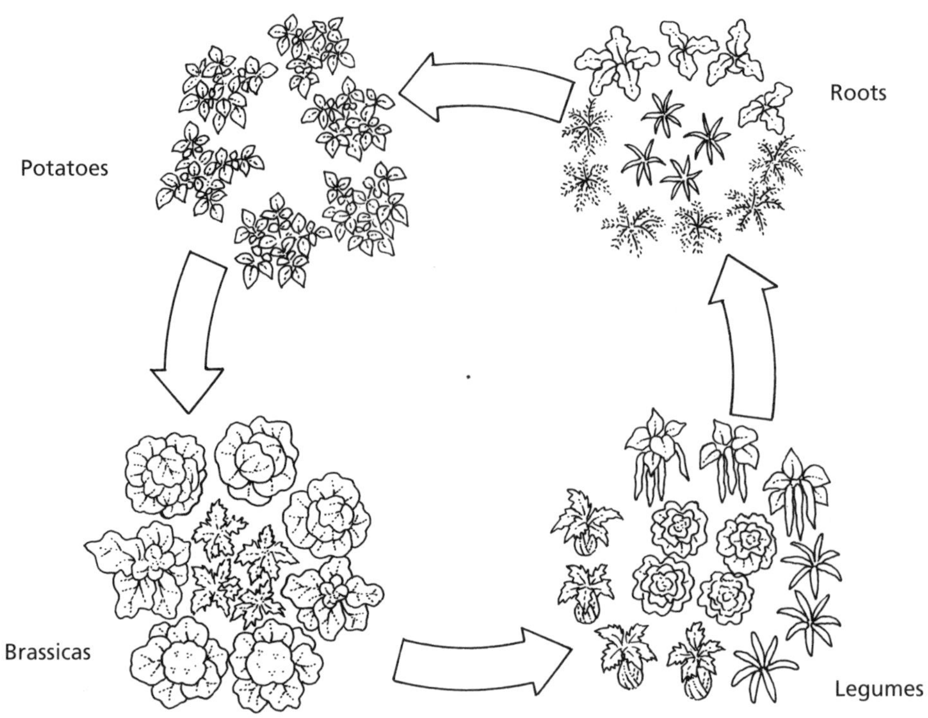

A four-year rotation for potatoes, roots, legumes and brassicas

are given manure (which is acid) and two are limed to make them alkaline. In alternate years each plot will be manured or limed (if necessary) so the pH should stay in balance.

Year 1 The soil is manured for the potatoes. They like a rich soil on the acid side. Potatoes are the traditional clearing crop for bad ground. They are leafy plants so they keep the weeds down. As there is quite a lot of excavation in potato growing, earthing up and harvesting the crop, soil-borne pests are exposed to the birds.

Year 2 The following year, the same soil will have about the right richness for growing roots (the carrot family). Roots prefer a light soil on the alkaline side, so add lime if the pH of the soil is low.

Year 3 The third year, the same patch is manured again for the legumes (peas and beans). They take nitrogen from the air and store it in their roots. The roots are left in the soil when the legumes are harvested. The nitrogen from them will be of great value to the leafy brassicas (the cabbage family) – the last in the succession.

Year 4 Brassicas like to be planted firmly so let the land settle over winter before planting. Lime it once more, if necessary, as their worst enemy – clubroot – is not as happy in alkaline soil as it is in an acid one. You don't have to be too strict about rotation for fast-growing brassicas.

Other families can join in the rotation.

Good partners are roots, beets and onions as they don't need rich soil, whereas legumes and cucurbits do. Lettuces and cut-and-come-again crops can be slotted in almost anywhere. The cardoons and artichokes, Good King Henry, rhubarb, seakale and sorrel are perennials so they need a semi-permanent position. Corn salad, New Zealand spinach, purslane and sweetcorn don't belong to the main family groups and can be slotted in where there is space.

Decide what you want to grow, then divide your crops into families so you know which should be in which bed. There might be some surprises. Who would think that tomatoes are part of the potato family or that artichokes belong to the lettuce clan?

MAKING YOUR BEDS

There are two schools of thought on making beds – to dig or not to dig. Digging is the time-honoured way of getting air into the soil, breaking up compaction and burying annual weeds. At the same time it gives you the opportunity to remove the roots of perennial weeds and to incorporate organic matter.

Single digging This is cultivating the topsoil only. Work a narrow strip in sections. Take out one spit (the depth of the spade) to make a trench. Put the soil that you have dug out at the other end of the strip. Dig out a second trench to the same width, get rid of any weeds, mix in organic matter and spade it into the first trench. Carry on in this way. When you get to the end, pile in the soil from the first trench.

Double digging This is the same process with the additional work of breaking up the subsoil as you go by plunging a fork into it, if necessary up to the hilt. This should drain off compacted soils and help deeper-rooting plants to penetrate to lower levels and find water for themselves. Generally, this is only done if there are drainage problems or when setting up the beds.

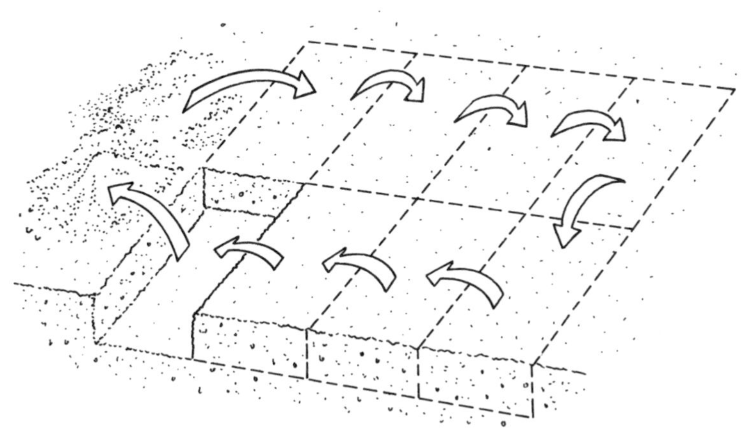

The best way to single dig is to work the section in strips

A FEW RULES OF DIGGING

- ✦ Never dig if the soil is frozen or waterlogged. If it sticks to your boots, it's too wet.
- ✦ Dig heavy soils in autumn and leave them in rough clods the size of a fist for the frost to get at them in winter.
- ✦ Dig light soils in spring so the winter rains won't leach out the nutrients.
- ✦ Keep the topsoil and the subsoil separate and put them back in their own layers.

Raised beds

Beds rise up naturally once you start adding organic matter on a regular basis. Though not necessary, it's a good idea to contain them to prevent leaching, define them and separate them from the paths. An edging of old planks is as good as anything. Drive a peg into each corner and nail the planks onto the outside of this. Logs sawn in half lengthways are a good alternative. Railway sleepers, long used in gardens, are ruled out as creosote is harmful.

The no-dig system

The principle of no-dig gardening is to leave the soil undisturbed. Every year you apply a thick mulch of well-rotted compost on top of the soil. This encourages earthworms, which build up fertility and improve the structure of the soil. As the soil stays moist under the mulch, the micro-organisms, which help to release nutrients to the plants, get an extra boost. The soil is protected from sun, wind, heavy rain and leaching and weed seeds are shaded out.

No-dig gardening has many adherents

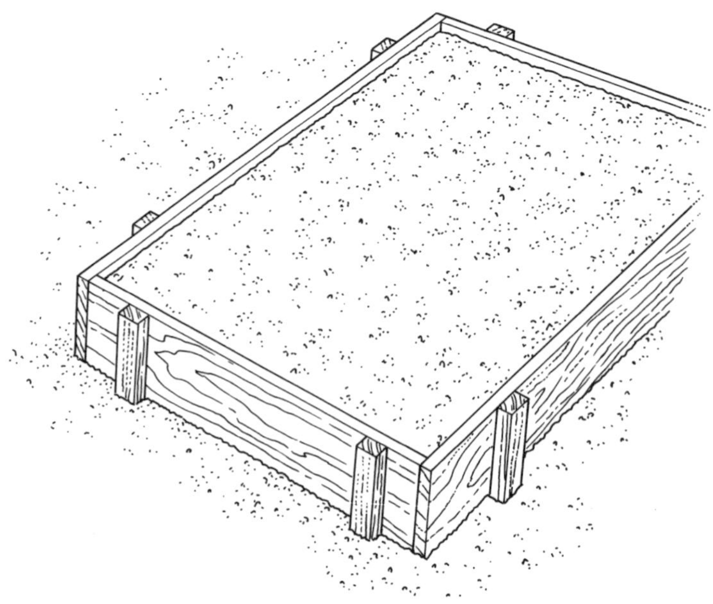

A raised bed made with planks and pegs

and works wonderfully well once you have good drainage and have got rid of any perennial weeds. My own feeling is to excavate the soil thoroughly once (if it needs it), then take up the no-dig system on raised beds – providing you have enough compost to make it work. The one argument against it is that any soil-borne pests will go unchecked, as they won't be turned over for the birds to find them.

GROWING FROM SEED

Buying seed

- ✦ Get hold of the free catalogues provided by seed merchants. They offer a far more interesting choice than the garden centres.
- ✦ For sure-fire results, buy top quality seed. It saves disappointment and wasting time.
- ✦ Look for disease resistance as it makes life much easier.
- ✦ Keep an eye out for the RHS Award of Garden Merit which is shown on seed packets. The award is given after extensive trials, to plants that do not need specialist growing conditions or care, have a good constitution and are widely available.
- ✦ Where possible buy organic seed. It is grown without man-made chemicals and on clean land. Most seed companies have an organic range and there are specialist organic seed merchants.
- ✦ Try some of the heritage varieties from seed libraries. Have a chat with your neighbours and find out which types work best in your area.

Plants that come from warm countries – tomatoes, aubergines, peppers and okra – need three or four months of sunshine to grow to maturity. They have to be started off in a warm greenhouse or in the home early in the year in Britain. If you only want a few, you can buy small plants all ready to go out in June.

Types of seed

F1 hybrids These are the result of the scientific crossing of two parent lines, resulting in reliable, uniform results. F1 stands for 'first filial generation'. With vegetables they concentrate on vigour, flavour and disease resistance. The criticisms of F1 seeds are that they are more expensive and that all the resulting plants mature at the same time – good for commercial purposes but perhaps not so desirable for the home grower. Of course, you can plant a few at a time to space out the crops. Don't save seed as there is no guarantee as to how the second generation will turn out.

Pelleted seed This has been devised to make sowing fine or irregular-shaped seed easier and more accurate.

Seed tapes These are designed for trouble-free perfect spacing to avoid thinning. The tapes are biodegradable.

Pre-germinated seed This is useful for tricky plants like cucumbers and melons which need Mediterranean warmth.

Primed or sprinter seed This has been germinated and then dried and must be sown within two months. It is usually more successful than untreated seed if you are

TESTING SEED FOR VIABILITY

Soak a few seeds overnight then put them on wet kitchen towel in a seed tray (or plate) covered with a plastic bag. Keep them at about 20°C (68°F). If all is well, they will start to sprout within three weeks.

planting early in the season or in adverse conditions. However, it is easy to pre-germinate them yourself.

Check that treated seeds have not had chemicals added.

Seed from the grocer In allotment circles there is a good success rate in growing from seed that has been dried for cooking. Notable examples are chickpeas, Egyptian red beans and other legumes. It's certainly worth a try if you are willing to take the risk. Check that the seeds are alive, or viable, before sowing as no seed keeps forever.

Your own seed Save your own seed (see page 172).

Sowing seed

Work out how many plants you want and add a failure margin as there might be a few duds. There is a UK statutory minimum level of viability for bought seed. This varies from plant to plant. While most have a near 100 per cent rate of viability, others are less predictable. Carrots only have 65 per cent and the tricky parsnip is let off altogether.

Seeds need warmth, moisture and air to germinate. Cold, waterlogged conditions are likely to be fatal. Avoid planting seeds too deeply – the seed carries within it all it needs to germinate but only has the power to grow a limited distance. It needs to reach the light before it can grow further. A general guide to planting depth is two-and-a-half times the size of the seed. Sow as sparsely as possible to save time thinning the seedlings later.

You need to decide whether to sow directly outside or to start them off under cover or in a nursery bed and transplant later.

GENERAL PRINCIPLES OF SEED SOWING

Read the directions on the packet carefully and follow the instructions on depth of planting, sun and shade considerations, cultivation tips and when to sow. Use common sense on the sowing times. You know your local weather best. You can often plant more closely than recommended if you want smaller vegetables.

It's a good idea to have a seed-sharing scheme with your neighbours as there may be far too many seeds in one packet for a single family – 500 or more quite often. Keep a few back, however, in case of bad luck or foul weather.

Sowing under cover

This has many advantages over sowing outside. You can begin earlier and provide the best conditions for a flying start. You can also protect seedlings from pests until they have built up strength and vigour. Brassicas, which recover well from being moved, are usually started off in this way. Tender plants like tomatoes or aubergines are also generally started off under cover and transplanted outside at a later stage as they need a longer summer than the British climate can provide.

A heated propagator, which provides the right level of heat from underneath, is a good investment. If you don't have one, you can manage very well with an airing cupboard or the top of a radiator kept at a low 20°C (68°F) and a light windowsill out of direct sun.

Choosing containers Almost any container is fine for growing seeds.

- ✦ Old tin cans, ice cream tubs and yogurt pots will do as long they are well scrubbed and have drainage.
- ✦ The traditional 10-cm (4-in) clay and plastic pots and seed trays work well. Clay is porous and so your seeds are less likely to suffer from over-watering, but it is less hygienic than plastic.
- ✦ If you are sowing seed big enough to handle and are buying new, go for the module or plug style of planters where each seed has its own place. The great advantage of these is that they pretty much eliminate any damage to the roots when transplanting.
- ✦ If you already have standard seed trays you can buy small square plastic pots designed to fit neatly into them. Good inventions are biodegradable products made from paper and alternatives to peat. When you are ready to transplant you plant them whole (pot and all), thereby completely avoiding transplanting shock. The modules can be broken off individually.
- ✦ You can also get biodegradable tubes, or root trainers, for plants with long roots such as runner beans. The tubes direct the roots downwards which will help the plant later to find water low in the soil. They can be opened up for transplanting with minimum disturbance and can be used time and again. The inner cardboard tubes from kitchen towel can be used in the same way.
- ✦ Polystyrene cell blocks with a presser board to pop out individual seedlings keep the seeds warm. Old polystyrene cups are the DIY version for seed warmth.

How to sow With a few exceptions, seeds germinate best in the dark. Fill your container with compost, leaving a couple

SEED COMPOSTS

Don't shortcut when it comes to compost. Use new seed compost, and potting or multipurpose compost for growing seeds. They contain everything that seedlings need, including nutrients, and are the right texture for moisture-retention and aeration. They are also sterile – free from weeds, pests and diseases.

of centimetres at the top for watering. Shake it from side to side to get a flat surface and firm it down gently. A firming board, or piece of timber cut to the size of the tray with a handle on the top, is useful for this. If you are using pots press down with the bottom of another.

Either water the compost thoroughly before sowing or sit the tray in water afterwards and the compost will soak it up by capillary action. You shouldn't need to water again until the seeds are up. While rainwater is generally best for plants it is safer to use sterile tap water for seedlings.

If the seeds are big enough to pick up by hand, make holes with a dibber (a pencil or chopstick will do the trick) and drop them in to the right depth. Fine seed should be scattered – carefully to get even distribution – on the top. The easiest way to do this is to have some in the palm of one hand and take a pinch at a time with the other. Mixing fine seed with a little silver sand helps to distribute the seed more evenly. Sieve over a thin layer of compost.

Cover the tray or container with polythene or a sheet of glass and put it in a warm dark place. Ideal temperatures for germination vary but few seeds like a temperature of more than 20°C (68°F). Look every day for signs of life. As soon as the first seed makes its appearance, put the seedlings in a light place but out of direct sunlight and take off the cover.

Caring for seedlings

Watering The most common problem with seedlings is 'damping off' – a fungus problem caused by wet or poor hygiene. Keep the seedlings just moist. Water slowly at the base so as not to disturb them, or let them soak up water from below by capillary action.

Light If you are growing them on a windowsill, they need to be turned every day to get even light all round. An alternative is to put them in a box lined with kitchen foil to reflect light.

Air Seedlings need good air circulation but keep them out of drafts.

Thinning If they become overcrowded you need to thin them out. Growing them too close together leads to competition for water and nutrients, and can cause weak growth, even disease. Water well before you start, then pull or ease out (an ice lolly stick is an effective tool) the surplus, leaving the strongest seedlings. If they are entangled, pinch off the leaves and stems of the ones you don't want at ground level.

Pricking out and potting on

When seedlings grow too big for their container or, if they are ready but the weather is not, you will need to prick out and pot on. This is best done young when the plants have two true leaves. These are the second set, the first leaf-like appearances being the seed leaves. Water well ahead of time and use a dibber to ease out the seedlings gently, taking as much compost with them as you can to avoid disturbing the roots. If you need to support the plant, hold the leaves, not the stem which is easily damaged.

If you have grown them singly in modules, just tip them out and plant in bigger pots to the same depth.

Sowing outdoors

Once the soil is warm, most plants can be sown outside. Some plants – notably root

TRANSPLANTING SEEDLINGS OUTSIDE

Acclimatize your plants for the world outside in stages. To start with, keep them in their pots under cloches or the cold frame, lifting off the cover by day and putting it back on at night. Protect them from fluctuations in temperature until they are tough enough to cope. The danger with transplanting is that young plants can get checked or suffer from transplanting shock due to root damage and drying out. To avoid this:

- ✦ Where possible grow them in plugs or modules.
- ✦ Work at a steady pace. Have the land raked to a fine tilth, watered and lined up for planting before you start. Roots dry out extremely fast out of the soil.
- ✦ Transplant them in the cool of evening or on a dull day.
- ✦ Water them well the day before and keep them well watered after until they perk up.

vegetables, cucumbers and sweetcorn – don't transplant well and are generally sown where they are to grow. Onions usually go straight out as they don't take up much space.

As a rule of thumb, the hardier plants – most brassicas, Oriental greens, peas, broad beans and radishes – can be sown early outside when the soil temperature at night has reached a constant 5°C (41°F) for a week.

Tender vegetables, French and runner beans, sweetcorn and tomatoes need a minimum 12°C (52°F), unlikely until June. Most lie between the two. Err on the cautious side however. It is better to plant later and grow faster.

Preparing a seed bed

Having a seed bed in top condition will save the effort of bringing the whole allotment up to scratch. If you have heavy, wet, clay soil, it is worth making a raised bed with good drainage for sowing seed. Ideally, the bed should be semi-prepared the autumn before, by digging and mixing in plenty of organic matter. Leave it to settle through winter.

A couple of weeks before you are ready to sow in spring, warm the soil by laying clear polythene on top. This will encourage any dormant weeds to germinate so you can dispose of them before you start.

If the bed is prepared in spring you may need to flatten it down by walking up and down on it. Do this when the soil is reasonably dry and don't stamp. You are aiming for a flat and reasonably firm surface not a compacted one. Break up lumps by bashing them with the back of a spade and rake the bed until you have a fine tilth – crumbly soil in small particles that seeds can push through effortlessly. The

Use cloches and the cold frame for early sowings outside. They make a huge difference to the temperature underneath

Avoid windy days if you are planting fine seed outside. One gust could blow the whole lot away or, worse still, into another part of your allotment.

way to do this is to keep the rake as near parallel to the soil as possible and patiently use a forward and backward movement from different directions. Sprinkle on some general-purpose fertilizer, if necessary.

How to sow outside

Mark out the space so that you will know where exactly your plants will be. Planting in rows or in a geometric pattern helps you to discern what is weed and what is seed as the little plants appear. Define each row (even if it is a straight line in a pattern) with a gardener's line, or string with pegs at each end, to make sure that it is straight. Choose a hoe of the desired width (or use a corner) to mark out the rows or drills to the right depth for the particular seeds. For smaller seeds a stick will do the job. Use a measuring stick to get equal spacing between rows or individual large seeds.

Parallel drills These are practical for beans, peas and other climbers that need support. If they run north/south they won't cast shade on each other. Climbers can also be grown up strings attached to a central pole though there will be more of a shade problem.

Blocks For lower-growing plants, try

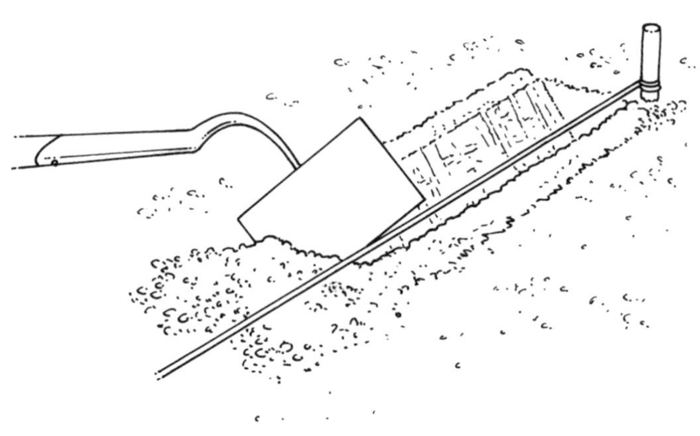

Making a straight drill along a string line

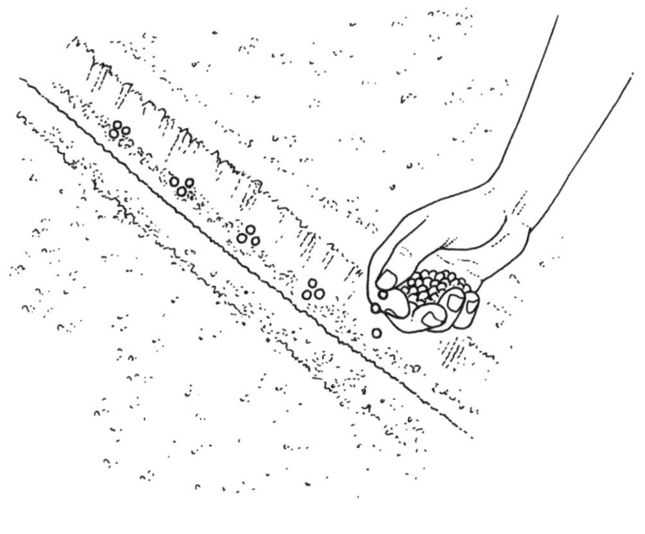

Station sowing

planting in blocks or squares. As the plants grow, their leaves will grow together, shading out weeds and keeping in moisture. They will also be easier to manage when netting. Still being in a pattern, it is just as easy to see where your plants are.

Station sowing This is used for vegetables that will be some distance apart. Three or four seeds are planted together. The strongest one is chosen and the remainder are thinned out. If carefully done, sometimes they can be transplanted.

Sowing seeds individually If you are planting large seeds, such as spinach, marrow or broad beans, quite far apart, the simplest method is to make individual holes with a dibber.

Broadcasting seed If you are planting in blocks with fine seed like carrot seed, you can broadcast it for speed then rake it carefully (so as not to bury it) in both directions. Outside you can use a little fine dry soil or horticultural sand as a 'spreading' agent to bulk up the seed.

Fluid sowing This is a modern commercial technique that gives surprisingly rapid and even results. It is particularly effective for tricky customers (like parsnips) and for getting a crop going (like sweetcorn) when the soil is a little too cold for it. The seeds are germinated without compost. Then they are mixed into a gluey medium, a carrier gel, and squeezed out like toothpaste onto prepared ground (see page 156).

VEGETATIVE PROPAGATION

Many perennials can be propagated by division – by separating the roots, snipping off offsets, chopping up tubers or taking cuttings. The results are faster than growing

FLUID SOWING AT HOME

The home method for fluid sowing is to line a plastic container with a few layers of wet (but not sopping) kitchen towel, lay the seeds evenly on it, put on the lid and keep it at 21°C (70°F). Check for progress regularly. You want to catch them at the stage when they have just germinated, still look embryonic and before the roots get too long. For the majority of vegetables the roots should only be about 5 mm (1/4 in) long. Rinse them carefully off the paper towel into a fine sieve.

Mix some fungicide-free wallpaper paste at half strength or some water-retaining granules (sold for container gardening) with water. When the mixture has thickened, add a few seeds, taking care how you handle them. Use tweezers or the tip of a plant label. If they sink, the paste isn't thick enough so adjust it as necessary. When you are satisfied, stir the seeds in.

Put the mixture into a plastic carrier bag and snip off one corner. Ooze them out evenly along the prepared drill (rather like icing a cake) and cover with compost in the normal way. Don't let the mixture dry out.

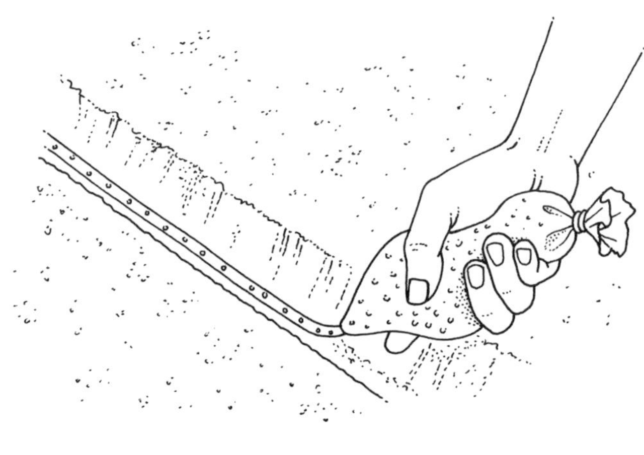

from seed, it is less work and the cost is zero.

Rootstock division This works well with asparagus once the plants are three or four years old. Dividing them will give the plants a new lease of life. Lift them when dormant in autumn, late winter or when they are just stirring in early spring. Divide up the roots so that each piece has a few buds and roots. Asparagus usually pulls apart quite easily, or you can use a sharp knife. Replant the divisions.

Dividing offsets Some plants – globe artichokes and cardoons, for example –

produce baby plants next to the rootstock. When they show themselves and have grown some roots in spring, carefully scrape away some soil and detach them with a clean sharp knife. Trim off the leaves bar one or two and replant firmly, making sure that the tip is above soil level. Keep well watered and warm under fleece until they are established.

Dividing tubers Jerusalem artichokes can be divided up to make more plants. Dig them up and cut the larger ones at the joints with a clean sharp knife, making sure there are healthy buds on each section. Then replant.

Root cuttings Seakale is generally grown from root cuttings, also known as thongs. Mark the spot where a healthy and mature seakale is growing before it dies down for winter. During its dormant period, dig it up, taking care not to damage the roots. Cut off a few of these on the outside, noting which end is which. Usually this is done with a straight cut at the top of the root and an angled one at the bottom of the cutting. Tie them in bundles of five or six of equal length and bury them in moist sand up to the neck the right way up. In spring they should begin to sprout.

Don't delay. Remove all but one bud and plant the individual roots upright with a dibber. The tops should be buried about 2.5 cm (1 in) below the soil surface. Keep moist and remove flower buds as they develop. You want them to put their energy into making new roots.

The best time to grow cut-and-come-again crops is in the cool of spring and autumn when they are less likely to bolt. Make sure the bed is completely free of weeds before sowing. If you are using the land to this level of intensity, it's important to keep it well fertilized and in top condition.

MAKING MAXIMUM USE OF SPACE

Catch cropping

There are some fast-growing plants that can be slotted into fallow land between main plantings. This makes good use of your best beds and keeps up weed-suppressing cover. For example, you may find you have a gap of a couple of months between harvesting winter crops and planting tender ones in summer.

Good catch crops include the quick-growing green manures, such as fenugreek, mustard, phacelia and buckwheat, as they will put goodness back into the soil (see page 142).

In terms of vegetables, the speediest growers are baby carrots, beetroot, radishes and the cut-and-come-again (CCA) crops – the gourmet baby salad leaves that cost a fortune in the shops. 'Salad Bowl' lettuce, cutting lettuce, leaf and seedling radish, curly endive, alfalfa, spinach, sugar loaf chicory, corn salad, and the Orientals (particularly mizuna and mibuna greens) grow fast.

Successional sowing The idea is to plant little and often. When the first sowing has germinated, sow a few more seeds. In this way you can keep a constant fresh supply in the amounts you need.

Intercropping

This makes use of the gaps between slow-growing plants that are spaced for the size they will be at maturity. Cauliflowers, Brussels sprouts, celeriac, parsnips, Hamburg parsley, salsify, scorzonera and purple sprouting broccoli will eventually take up a lot of space but, until they do, there is room between them for speedy crops. However, don't let them overlap too much and compete with the main crop.

Undercropping

This is making use of the space under tall vegetables. Sweetcorn, garlic, Brussels sprouts and the climbers (beans, cucumbers, tomatoes or squashes grown up canes) are ideal for the purpose.

Strip cropping

This is an economic and labour-saving method of growing vegetables, using parallel rows of plants and moving the cloches from one row to the next. The principle is to plant early, middle and late crops which need cloche cover at different times. So through winter you could have hardy vegetables under the cloches. In late spring these would be left uncovered and spring sowings would be planted in the next row under the same cloches. By summer the spring sowings would prefer to be out in the fresh air and cloches could be used for tender vegetables – tomatoes, aubergines or peppers.

SPACING AND MINI VEG

When you buy seed, you are given recommended planting distances. However, in many cases the distances can be manipulated depending on whether you want large or small vegetables. If you want small onions for pickling, baby carrots or miniature cauliflowers, plant them closer together than recommended.

The Institute of Horticultural Research has run tests on different planting distances. If Brussels sprouts are to be allowed to grow to full height and treated as a cut-and-come-again crop, the full recommended distance of 90x90 cm (3x3 ft) is about right, allowing access for picking and room for hoeing. However, if you want a bulk supply of small sprouts for freezing, plant them closer at 50x50 cm (20x20 in) as commercial growers do. Remove the growing point and small leaves of the plant when the biggest sprouts are the size of your little fingernail. Then a single succulent harvest is taken when the bottom sprouts are just past their best.

WATERING

On an allotment, watering is a time-consuming business, generally done by hand from a stand pipe or water butt. Quite apart from conservation issues, you want to keep it to a minimum.

Conserving water

✦ Maximize the water-holding capacity

of the soil by adding lots of well-rotted organic matter on an annual basis.

- ✦ Eliminate the competition for water by chasing after weeds.
- ✦ Mulch after a good rainfall to keep moisture in.
- ✦ Conserve water by making a shallow moat or reservoir around plants.
- ✦ Collect as much as you can in water butts. Put in guttering and downpipes to collect it from the roof of your shed. Keep the butts covered to prevent evaporation and to stop them getting clogged with leaves.
- ✦ Sink a flower pot or section of pipe into the ground to get water down to the roots of thirsty plants, like courgettes or cucumbers.
- ✦ Check the moisture content of the soil by digging a hole with a trowel. If it is damp 23–30 cm (9–12 in) deep, the roots of established plants will have access to moisture. This doesn't apply to seedlings.
- ✦ Water in the cool of the evening or the crack of dawn to avoid rapid evaporation.

When to water

You can manipulate the performance of plants by timing when and how much water you give them within their growth cycles. Research show that the general response to watering is leaf growth – exactly what you want for leafy vegetables like cabbage or lettuce.

It is not what you want, however, for peas and beans. It is better to keep them on the dry side until they start to flower. At this point the root activity slows down and the plants could do with some help. A thorough watering will produce more flowers, hence more peas and beans. Potatoes profit from a good dousing when they are at the 'marble' stage.

As a rough guide, all plants need sufficient water when young until their roots establish. Vegetables grown for their leaves should never be allowed to dry out. Fast-growing plants, like courgettes and marrows, need plenty of water right through their short growing season. On the other hand, overwatering can produce a lot of soft growth attractive to pests, and make plants susceptible to cold or rotting off. Plants grown for their roots, pods and fruits need steady but not excessive water until the flowers, fruits or roots form.

How to water

When watering, make it thorough – 11 litres per square metre or 2 gallons per square yard. The water should soak to the lower depths of the soil to encourage the roots to grow down to seek it. If you just sprinkle water, the roots will look no further and can dry out as soon as you turn your back.

Water brassicas at the base to avoid the

Transplants are particularly vulnerable to drought. Water before and after transplanting. This is the one occasion when you should water little and often (every day or even twice a day in dry weather) carefully at the base of the plant.

> Occasionally, trace elements or minerals might be lacking in the soil. Iron deficiency on very alkaline soils and phospherus deficiency on very acid soils is not uncommon. If you are concerned, it is best to send off for a full soil analysis to pinpoint any deficiency.

damp conditions that can encourage fungal disease. For cold-sensitive plants, such as tomatoes, use water from the water butt.

Seeds and water

Dribble water into the seed drill before planting. Watering from overhead after the seeds have been sown can make the soil cap, or form a crust, making it difficult for the seeds to push through. Fine seed can be washed away. If you need to water after the seedlings emerge, use the finest rose on the watering can or let them soak it up by capillary action.

FEEDING

With high soil fertility you shouldn't need to feed much, as long as you keep up the regime of digging in or laying on a good quantity of rotted manure or compost on an annual basis. The Henry Doubleday Research Association, the organic organization, says that best practice is no fertilizer at all. It is a supplement, not a replacement, for bulky organic materials.

However, as it takes a few years to build up a rich and fertile soil, you may need to add fertilizers to boost the major elements for plant growth – nitrogen, phosphorus and potassium (generally known as NPK).

Other reasons for fertilizing are when your crops need a boost – hungry ones like tomatoes still indoors because of bad weather, or those that have overwintered outside and are looking a bit ragged. The exact amount and which type of feed depends on the plants, whether you want to improve the fruit and flower potential, or put in leaf growth.

How to feed

Use organic fertilizers which come from plant, animal and mineral origin found naturally. They will not harm the soil or the populations of micro-organisms within it whereas the man-made alternatives do. Standard practice, used until the soil is fully fertile, is a top-dressing of an all-round fertilizer like blood, fish and bone, or seaweed meal prior to planting. Liquid fertilizers are used to perk up plants in containers where nutrients may be running out or for a quick boost for failing specimens.

Major elements – macronutrients

- ✦ Nitrogen (N) encourages leaf and shoot growth. A component of chlorophyll, it gives plants their greenness. Too little and the plants will be stunted and the leaves pale; too much will produce sappy growth which will attract pests and collapse as soon as the first frosts arrive.
- ✦ Phosphorus (P) (or phosphate) encourages healthy growth throughout the plant, especially the roots. Only small quantities are needed. A deficiency will

show as stunted growth and a purple or red discoloration on the older leaves first.

- ✦ Potassium (K) (or potash) is associated with the size and quality of flowers and fruit. It toughens up the plants to protect them against pests and disease. A deficiency will show up as small fruit and flowers, also yellowing or browning of the leaves.
- ✦ Magnesium (MG) Like nitrogen, magnesium is a constituent of chlorophyll, the greening agent. A deficiency is usually a sign of insufficient organic matter in the soil. Magnesium deficiency causes chlorosis. The symptom is yellowing of the leaves starting between the veins.
- ✦ Calcium (Ca) helps in the manufacture of protein.
- ✦ Sulphur (S) is part of plant protein and also helps to form chlorophyll. It is unusual to find it missing in places where plenty of organic matter is added. The same goes for the trace elements.

Trace elements – micronutrients

- ✦ Manganese (Mn) is another player in the formation of chlorophyll and protein. A deficiency will show as stunting and yellowing of the younger leaves.
- ✦ Iron (Fe) plays much the same role as magnesium, though it's only needed in tiny quantities. A deficiency is more common on alkaline soils.
- ✦ Copper (Cu) and Zinc (Zn) activate enzymes.
- ✦ Boron (B) is an important element throughout the growing tissue. A lack can cause corkiness in fruit and root crops.
- ✦ Molybdenum (Mb) helps to produce protein.
- ✦ Oxygen, carbon and hydrogen are taken up from the air, sunlight and water.

Organic fertilizers

- ✦ Blood, fish and bone meal is a balanced NPK all-round fertilizer.
- ✦ Blood meal provides nitrogen and is used as a quick tonic for overwintered crops in spring.
- ✦ Bone meal is high in phosphate for root growth and is useful when planting shrubs and trees.
- ✦ Dried manures have all the trace elements but are low on NPK.
- ✦ Epsom salts are a soluble form of magnesium alone.
- ✦ Fish meal contains nitrogen and phosphate.
- ✦ Hoof and horn is high in nitrogen and works as a slow release. It needs to break down, so apply it to the soil a week or so before planting.
- ✦ Rock phosphate is a good alternative to bone meal for dog owners. It promotes rooting, particularly for shrubs and trees.
- ✦ Rock potash is a useful source for potash alone – a benefit to poor and light soils. It works as a slow release so is a good booster for fruit vegetables like tomatoes.
- ✦ Seaweed meal is an excellent slow-release all-round tonic with all the trace elements. It contains cytokinins, hormones which promote photosynthesis and protein production, apart

COMFREY FERTILIZER

Comfrey is a superb all-round fertilizer, though some people can't stand it as it stinks! The leaves can be used as an activator on the compost heap, thrown down into a potato trench or laid as mulch around crops, though as it is alkaline do not use it on chalky soils. Its long roots draw up potassium from the subsoil. Be careful where you plant it though – the roots are so deep that it is almost impossible to dig out.

Site it where it can stay for twenty years or more or persuade your committee to plant a patch. The non-invasive type is Russian comfrey (*Symphytum x uplandicum* 'Blocking 14'). As it rarely sets seed, the best way to start off is to buy some root offsets or get some off a neighbour. Plant in spring, or early autumn, though any time will do except deepest winter when it is dormant.

Incorporate some well-rotted compost and clear weeds, then plant 60–90 cm (2–3 ft) apart with the growing tips just below the surface.

Don't harvest in the first year but remove the flowering stems and go easy in the second year while the plant builds up strength. After this, comfrey can be cut back three or four times a year with shears when the leaves are about 60 cm (24 in) high.

Nettles make a good general tonic, though with lower phosphate levels than comfrey. Like comfrey, they are not suitable for alkaline soils. Nettles are richest in nutrients during the spring months.

from helping to create humus in the soil.

- Wood ash from your own bonfire (if allowed) is high in potassium and has some phosphate, though quantities of each depend on the type of timber.

Liquid fertilizers These don't stay long in the soil. Seaweed extract is an excellent all-round tonic for promoting strong growth and supplies the full range of trace elements. It is easy to make your own first-class liquid fertilizers out of comfrey, nettles, or sheep, cow, horse or goat manure. The technique is to tie it up in a hessian sack and leave it to steep in a barrel of water over a couple of weeks.

PROTECTING CROPS FROM COLD

Windbreaks

While plants need plenty of air circulation to thrive, they suffer badly in strong cold winds. If you are on a blustery, chilly site it's a good idea to set up some windbreaks. A good windbreak will work on a diminishing scale for a distance up to ten times its height, cutting down the wind by 50 per cent or more. So a 1.5-m (5-ft) high

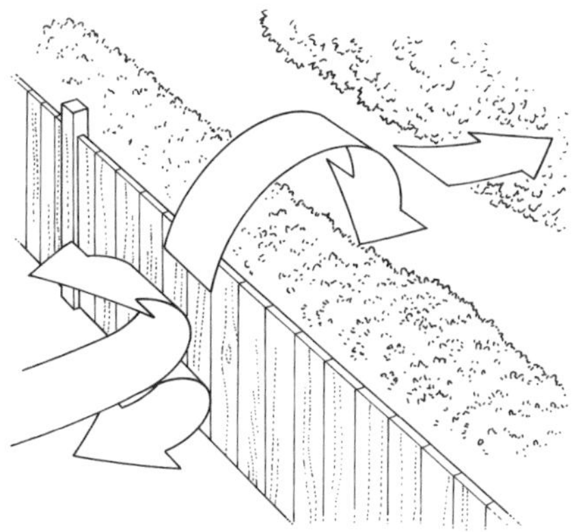

A solid windbreak creates turbulence while a permeable one filters the wind

If you have a full plot on a windy site, keep the more vulnerable plants at the end of the allotment near a tall windbreak, and erect lower localized windbreaks on the windward side of crops with sacking, mesh or fleece stretched across low stakes.

windbreak would cover the area up to 15 m (50 ft) in front of it. To protect the whole of a full-sized allotment, therefore, you might need two or three windbreaks going down in scale towards the hardier plants.

The most effective windbreaks filter the wind. It is fruitless to try to block it with a solid structure. If wind meets a dead end it will turn into turbulence, creating icy eddies on both sides.

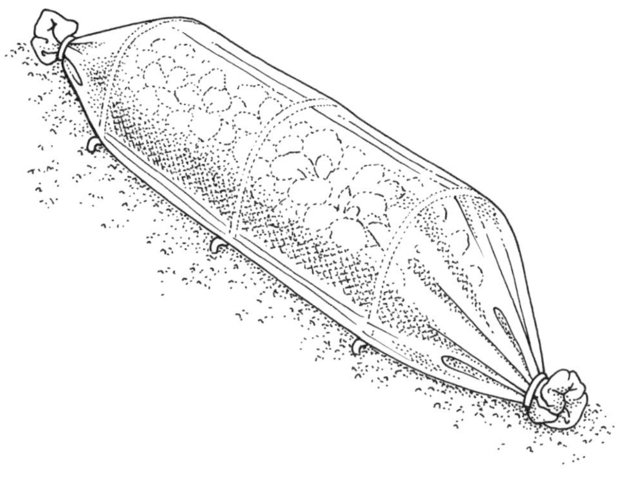

A portable mini cloche made from hoops and fine mesh

A good deciduous (or mixed evergreen and deciduous) hedge is very effective and has the additional bonus of making habitats for wildlife. Remember to plant well away from it as it will take up nutrients and moisture from the soil.

Wooden fences with gaps, lathe and wattle, or wire or plastic netting stretched across stout posts also work well and give you a structure for climbing plants.

Localized protection

Developments in plastics have proved highly efficient for bringing on and protecting plants from cold, wind and pests. Polytunnels, plastic cloches, floating mulches of polythene, horticultural fleece and mesh have transformed the gardening scene.

The non-porous types of cover, like polythene, come into their own in spring, autumn and winter. They are less good in the heat of summer when the atmosphere underneath can get too hot and humid and foster fungal disease. Plants which have been grown in a warm atmosphere need to be watched, aired and hardened off with extreme care.

WHAT PLANTS NEED

Cool climate crops, such as cabbages, cauliflowers, radishes, sprouts and turnips, will germinate when the temperature has reached a consistent minimum of 5°C (41°F).

Carrots and parsnips need a slightly warmer 6°C (43°F). By warming the soil and using protective cover, it is reckoned that you can keep the temperature above 6°C (43°F) for some extra three to four weeks at each end of the season.

Greenhouses

The greenhouse is an expensive structure and difficult to heat without electricity. Quite a few enthusiastic allotment gardeners have a heated greenhouse at home to raise young plants. More common on allotments is a cold greenhouse. If you are lucky enough to have one you will find it very useful for extending the season, for propagating out of the rain, and for growing tender crops and winter salads.

MAKING A COLD FRAME

When making a frame, consider the size of the plants that you will want to grow. You don't want your plants to become weak and leggy by over-reaching themselves to get to the light. If you plan to grow tomatoes, you would need a frame 1 m (3 ft 4 in) high but for salad crops you will need only a fraction of that. A selection of sizes would be handy.

You can buy kits to make cold frames. These are usually made of aluminium, steel or timber frame and rigid see-through PVC or other synthetic materials. An excellent version can be made from an old wooden drawer on the tilt with a window frame on top.

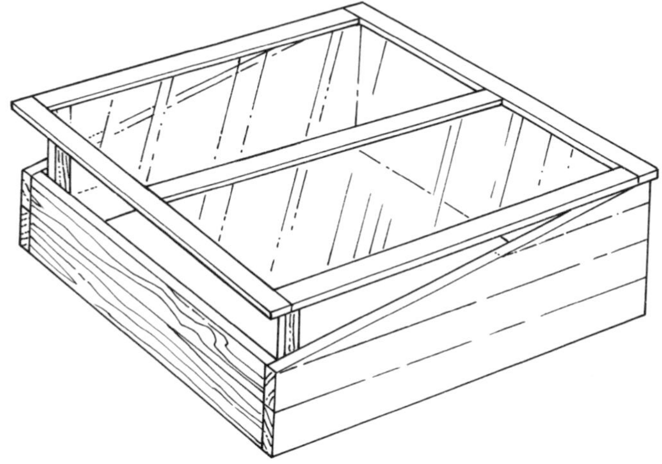

Coldframes of various sizes are invaluable for starting off crops and for cultivating tender ones

Polytunnels

A walk-in polytunnel is occasionally seen on allotments. It is believed to extend the season by six weeks in spring and four in autumn. It creates a near wind-free environment and works in much the same way as the cold greenhouse as the temperature and airflow need to be controlled manually.

Cold frames

The cold frame, which can be knocked up by a DIY handymen, is a must for hardening off seedlings, growing early and late salad crops and successional quick-growing crops and as a nursery bed for brassicas. In summer it's a good place to grow tender crops like ridge cucumbers, aubergines, peppers or tomatoes. Generally made with brick or wooden sides, it has a sloping glass or plastic roof that can be removed or lifted up for ventilation.

Hotbeds

If you have access to fresh stable manure you could try a hot bed. It's an adaptation of the cold frame with the advantage of providing bottom heat like an outdoor propagator.Victorian gardeners made great use of them on large estates for out-of-season crops and early delicacies for the house. It's a great idea for the allotment, especially for those who don't have the luxury of a greenhouse or propagator. It makes an interesting challenge, is free (unless you pay for the manure), low-tech and energy saving.

What you need You need a cold frame lid with glass or plastic windows. It could be made up from old window frames.The other essential is a good heap of stable or farmyard manure with straw in it – so fresh that it steams in the cold. If you find you haven't got enough, mix it with up to an equal quantity of newly collected leaves. Victorian writers recommend oak or beech, but almost any will do.

An adaptation of the hotbed is to use your existing compost heap. Square it off, add a layer of potting compost and cover it with a frame. Make holes for potting compost and sow your seed. Cucumbers and melons will thrive in it.

The bigger the heap, the higher the temperature you will achieve. To get a hothouse temperature of 27°C (80°F) you would need a heap 1.2–1.5m (4–5 ft) square once it has settled down. However, you can get useful warmth from a hotbed half that size. Turn the manure and straw at least once (some say twice or three times) a day, shaking the fork to break it up and get it well mixed. After a about week it will be ready to use.

Making the hotbed Decide whether you want to dig a pit or build it on the ground. A pit will keep things tidier and give some insulation. If you have cold wet soil, though, it's better to make it on the surface. If you go for the pit method dig down about 30 cm (12 in), reserving the top soil. If you are building it on the surface just remove the topsoil and make a flat base. The dimensions of the heap will depend not only on how much manure you have, but on the size of the frame. Allow for a good border, about 30 cm (12 in), around the frame for stability.

Pile the manure into the pit, beating it down so that it will stay firm and flat. Aim for a compact, squared-off pile. Leave for a few more days for the 'rank steam' to escape and for the heap to settle. Add a 10-cm (4-in) layer of the reserved top soil. Check the soil temperature – if it is over 27°C (80°F) wait a few days until it cools down. Cover the heap with the cold frame, leaving a little chink for gases to escape.

THE BENEFITS OF CLOCHES

- Cloches can be used to warm the soil by a vital degree or two if put on two weeks before planting or sowing.
- They will protect crops from birds, rabbits and insect pests.
- Cloches keep hardy crops warm through the worst of winter and protect half-hardies from passing frost and wind.
- Herbs and salads are given a longer season into early winter if covered with cloches and a racing start in spring.
- Another benefit is that they keep the soil on the dry side, discouraging fungal disease, slugs and snails. Rainwater, which falls each side, will still find its way to the plant roots.

Cloches

Cloches are portable cold frames which come in many shapes and sizes. The simplest is a plastic bottle with the bottom cut off (a bread knife will slice it off with ease) and the top removed for ventilation. Polytunnel cloches can be made from wire hoops covered in polythene, or insect netting with the edges buried or weighed down with stones.

Cloches can be bought in kit form or devised. There are dozens of different proprietary brands and types. Polycarbonate sheeting has more or less taken over from glass – a safer option if there are children about. Sheets balanced against each other into an A-shape are handy to cover a row.

Choosing a cloche Keep in mind that cloches need to be lifted for watering so they shouldn't be too heavy or awkward.

- ✦ They should let in plenty of light, especially in winter.
- ✦ They should be easy to open up for ventilation on hot days and to close to block out wind on cold ones.
- ✦ They need to be stable yet easy to move and assemble.

Horticultural fleece

This is a synthetic fabric (polypropylene), so light that plants barely notice that it is there. It provides warmth while letting in light, air and water so crops can be grown to maturity under it while being protected against pests.

If there is no fleece to hand and a frosty night is imminent, straw or newspapers under chicken wire will help, or any netting you can find – even old net curtains will do.

Fleece can be used to cover up crops throughout their growth. This is known as a floating mulch and speeds up growth quite dramatically. Make sure you've got rid of weeds as they will sprout underneath it with equal vigour.

When covering the crop with fleece, allow a little slackness for growth. The usual

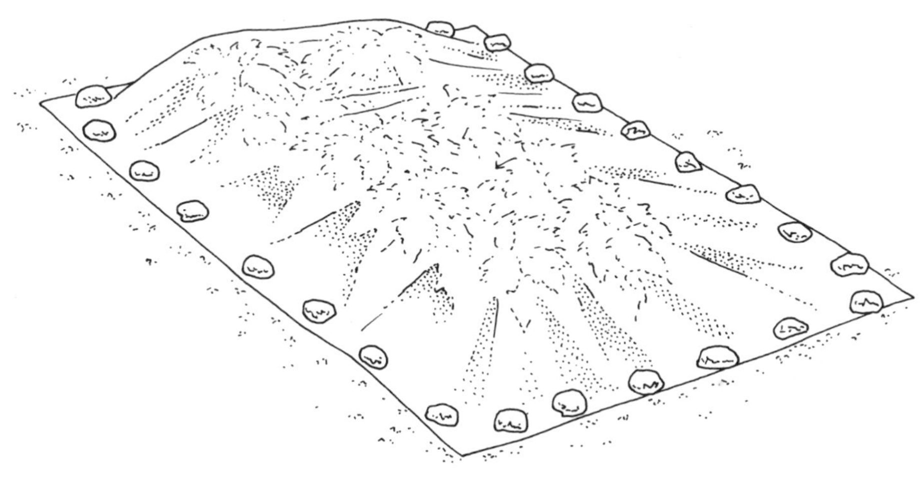

A floating mulch weighted down with stones

Polystyrene is a good insulator. A sheet under your seed trays will help to prevent cold creeping up from underfoot. A box made of polystyrene sheets, glued together with a see-through top, can make a cosy environment for plants in a cold greenhouse in winter.

method is to bury the edges so that you can let it out as the plants develop. If you have raised beds with wooden edging, the fleece can be stapled to the sides.

It's very useful to keep a roll of fleece to hand to throw over vulnerable plants if frost threatens. Its one disadvantage is that it will eventually get dirty and tear.

Other floating mulches

Mesh fabrics or perforated transparent film are made of plastic. They don't keep the plants as warm as fleece but are just as effective against pests and they last longer, especially if they are UV treated. As they let in more air than horticultural fleece, diseases are less likely to develop.

Use these products like fleece as a floating mulch. Mesh comes in various gauges; the smallest will keep out tiny pests like flea beetles. Get the right size for the particular pest you want to keep out.

Basic polythene sheeting is useful for warming large areas of soil by a degree or two. If you have any farming friends, ask them to let you have their old silage bags.

PLANT SUPPORTS

Climbing vegetables, like beans, are usually grown in rows of criss-cross poles

Supports for beans – the wigwam, cross poles and strings

CANE TOPS

Sharp cane tops are a hazard to eyes, so make sure to cover them. Some people swear by small plastic drinks bottles, others by film canisters or old tennis balls.

or on wigwams. Peas will scramble and cling onto twiggy sticks pushed into the ground, or up netting hitched between posts. Netting can be any lucky find, from an old fisherman's net left on the beach to an old tennis net. Some people use chicken wire and burn off the dried debris at the end of the season.

BLANCHING AND FORCING

Forcing is the way to get your own out-of-season crops by bringing the maturation date forwards. Chicory and seakale can be forced for winter eating. Blanching removes bitterness by excluding light.

Chicory Dig up the chicons in autumn, trim off the outside leaves and replant in a container (or box) in peat substitute. Keep them in a warm dark place. They should have grown to some 20 cm (8 in) and be ready to eat in a few weeks. To force outside in autumn, cover the plants with a flower pot, raised a little on crocks or tiles to let the air in, and block the hole. The 'Witloof' varieties are the best for forcing.

Seakale Tidy up the plant once it has died back in early winter by removing the old leaves, put on a layer of straw for warmth and cover it in a bin, bucket or forcing pot wrapped round with straw to keep it warm. Allow about 40 cm (16 in) for growth. It will be ready in two to three months.

Radiccio This can have the bitterness removed by placing a saucer over the heart for a week or so before harvesting.

Leeks and celery These can be whitened, either by earthing up, or wrapping the stems with cardboard, newspaper or roofing felt as they develop.

THE HARVEST

Through the height of summer, harvesting crops at their peak becomes a daily task, particularly catching the prolific fruiters like runner and French beans, tomatoes and courgettes. On every visit there will be the outer leaves of the pick-and-come-again crops to collect and something new to take home for supper. Harvesting becomes a race against the plants becoming coarse or bolting. Turn your back in June, July and August and the crops will become overblown and your courgettes will have turned into giant marrows! If you are going away, ask someone to pick for you.

It is as well to get organized before you start the big autumn harvest, laying in storage jars, freezer containers, crates, sacks, sand and whatever else you will need. Produce begins to deteriorate as soon as it is harvested, so aim to be fast and efficient.

GOLDEN RULES FOR HARVESTING

- ✦ Avoid damage. Only store perfect specimens – broken tissue invites fungus and bacteria and rough handling can lead to bruising.
- ✦ Use a clean sharp knife for any trimming. A clean cut will repair quickly.
- ✦ Cool down the crops as soon as possible after lifting. The ideal time to harvest is early morning on a cool day, moving the crops from the ground, to the shade of a tree, to the shed or fridge with due speed.
- ✦ Once stored check regularly. Remove any that start to rot to save the rest of the crop.

As autumn approaches, make a plan for the remaining produce. All frost-tender vegetables need to be brought in and stored in one way or another. Ripe tomatoes can be dried in the sun, made into ratatouille with the remaining courgettes and aubergines, or be the base for pasta sauces or juices. Those that haven't ripened as the days shorten make good green chutney.

In the same way, other vegetables and fruits can be dried, bottled, pickled and preserved and much can go into the freezer. To prevent frozen produce sticking together in a lump, freeze it spread out on a tray before putting it into bags.

To store or not to store

Only store vegetables that won't keep better in the ground. Most roots – carrots, parsnips, beetroot, swede, turnips and leeks – stay freshest where they grow. Growth slows down in the cold to a gentle ticking over and they will have the benefits of natural (free) refrigeration. You can give them some extra protection against frost by earthing them up, putting on cloches or covering them thickly with straw or leaves. Don't do this until the last minute (December in most places) as the additional warmth could encourage growth. Choose winter types as they grow more slowly and cope better with the cold. Carrots and parsnips generally survive winter well – parsnips becoming sweeter for the experience. Beetroot, swedes and turnips are usually dug up and stored or eaten before the end of winter as, if left too long, they can become coarse and fibrous.

However, if you have a pest problem, or if the soil temperature in your allotment is likely to drop a few degrees below freezing, it would be wise to lift and store them in autumn.

The clamp

A traditional way to store root vegetables is in a clamp. It's practical for the allotment as there is usually more room outside in winter than in the shed. A clamp will keep the vegetables fresher for longer than if you store them in sacks or crates.

First decide whether to make a cone or a narrow rectangular shape like a roof ridge. If you will want to raid it bit by bit, the long thin clamp is more practical than a cone as you can seal it up again as you go without fear of it collapsing.

Make a base layer of sand or sandy soil

with some wood ash or bran around the edge to deter slugs. Choose a fine day to lift the vegetables. Lay them out on the ground for an hour or two to dry off in the sun and air, turning them over when the first side is dry. Sort through and remove any damaged or diseased ones. This is important as one bad specimen could wreck your entire hoard. Twist the tops off turnips or carrots to about 2.5 cm (1 in) and rub off excess mud.

Building the clamp Build a 'castle' of the vegetables, with the largest at the bottom and the smallest at the top aiming for a pitch of 45 degrees. Cover generously with vertical layers of straw, dried grass or reeds. Wheat straw is the easiest to work with as it's straight and rigid. The straw keeps soil off the vegetables and protects them from frost. Tie it at the top to make a top knot. It will poke out of the finished clamp like a chimney and act as an air vent. For a long thin clamp you will need several of these, spaced a few feet apart.

Pack a 15–20 cm (6–8 in) layer of soil around the structure, patting it down firmly with your hands. Work from the bottom, tier by tier, pushing soil into cracks and hollows. End by compacting it with the back of a spade.

Removing the vegetables When you want to get your vegetables out, make a hole by chipping away at the side near the base. If there are any signs of structural damage, pack in more soil. Don't open the clamp in frosty weather.

Potatoes are particularly frost-sensitive, so if you want a good supply for the New Year, get enough out to see you through. If the weather becomes icy, close up the chimney but don't forget to open it again as soon as possible. As long as the soil layer is thick enough, you shouldn't have any problem with rodents. However, if you do, cover the clamp with galvanized wire mesh.

Most vegetables are pollinated by insects, including kale, cabbage, broad beans and leeks. A few are pollinated by wind, including beetroot, sweetcorn and spinach.

COLLECTING SEED

Collecting and growing from your own seed is great fun and makes you feel like a real professional. It saves money, conserves rare breeds and encourages local diversity. Always select your best specimens. For safety's sake remove any nearby plants that look less than healthy and any other varieties of the same species as you don't want cross-breeding with the wrong plant.

The plant's biological clock

Before you start you need to gather a few facts. You need to know when the plant will set seed. The vast majority of vegetables are annuals, so if sown in spring they will set seed by autumn. Others, including parsnips, carrots, onions and Brussels sprouts, are biennials and won't flower until the following spring. They need to be left in the ground over winter. The more tender vegetables, like carrots, that might not survive the experience, are dug up and stored in moist sand in a cool shed or can be potted up to be replanted the following spring. The perennials are propagated more easily from cuttings than seed.

Keeping the strain pure

Plants are divided into self-pollinators and promiscuous cross-pollinators and others that do both. The self-pollinators, like French beans, are the easiest to keep pure. Peas self-pollinate before the flowers open, making them very reliable.

The vast majority of vegetables are cross-breeders with their own kind. The large cabbage family of brassicas will cross-pollinate freely and indiscriminately with each other. If they are crossed with a different variety or a different member of the same species (it could be a weed) you'll end up with an interesting mongrel – a cross of celery and celeriac, or sugar beet with beetroot, for example.

Commercial crops are prevented from cross-pollinating by being kept well apart. In an allotment situation, it is impossible to isolate plants from each other by distance, but they can be kept 'pure' by netting them with fleece or fine mesh.

If you want just a few seeds, individual flowers can be tied up in fleece bags. Put the bags loosely over both male and female flowers, possibly ten per plant, when the flowers are on the verge of opening. Wrap cotton wool around the opening to make sure that even the smallest insect is kept out. Tape it securely. If you want a lot of seed, the whole crop can be netted in a cage. If the plants are not self-pollinators, you will need to hand pollinate them.

Hand pollination When the flowers open, take off the bag and carefully pull off the petals to get to the pollen on the male flower. With a small paintbrush, tickle out some pollen and take the bag off the female plant. Brush the pollen onto the stigma of the female flower. Alternatively, cut off the male's flower, pull off the petals and brush the anthers onto the female stigma. Cover without delay before any insects arrive. Repeat this over the plant and any others of the same species and variety nearby. You want a good cross-section to avoid inbreeding. The pollen from one male flower will pollinate quite a few female ones. When the flowers wilt, take the bag off and wait to see if the fruit will form.

SEPARATING SEED FROM CHAFF

For fine seed, like carrot, work over a colander (possibly several going down in size) so that the seed falls through. Heavy seed can be separated by winnowing with a hair dryer. If you have a lot you can thresh it by putting the seeds in a sack or bag and treading on it. Check through the seeds for signs of mouldiness, disease or maggots.

Harvesting seed

Vegetables are either dry-seeded or soft-fruiting.

Dry-seeded When the seed in seedheads, pods or capsules is ripe it is usually so dry that it rattles. It is best to let nature choose the moment. Tie brown paper bags over them when you deem them to be nearly ripe to catch the seeds. If the weather turns and the seeds have already gone brown, cut down or dig up the whole plant and hang it upside down in a dry shed.

Soft-fruiting vegetables Seeds of tomatoes, squashes and cucumbers are harvested when the fruits are perfectly ripe for eating. You can store them a little longer to ripen further. Give them a good wash to get rid of the pulp and the gel. If you are being thorough, soak them for a day in water with a teaspoon of washing soda. You can tell that the gel is off when you see the debris floating on the top. Rinse and spread them out to dry quickly before they start to germinate or go mouldy. Put them in a warm place, out of direct sunshine, and turn them over every so often. Some speed things along by using a hair dryer set on cool.

Drying

A little further drying will give your seeds greater shelf life. Spread them out on a tray and put them in the airing cupboard or on a windowsill at a maximum temperature of 35°C (95°F). They are ready to store when they are so dry that they shatter if hit with a hammer.

STORING SEED

Seed deteriorates quickly if not stored well. Keep commercial seed sealed in the original packets or your own in paper bags in an airtight jar in the fridge or a chilly cellar. Buy some silica gel from the chemist and put this in the jar to keep them bone dry. Test for viability if you are planning to sow them the following year (see page 150). Keeping in mind that less will germinate outside than in, you really want a minimum of 60 per cent germination rate when you do the test.

CHAPTER 4

PLANT PROBLEMS

ORGANIC PRINCIPLES

Controlling pests and diseases organically is not as simple as changing from man-made sprays to organic ones. It is far more interesting. It is about creating a partnership with nature. It comes down to good husbandry, knowledge, observation and using your wits.

With netting, crop covers, traps, camouflage, timing and an increasing choice of disease-resistant varieties, you have a powerful armoury at your disposal to control pests and diseases – without chemicals. It is an interesting fact that the top growers for exhibition are 95 per cent organic. They avoid chemicals as they know that they will have an adverse effect on the quality of their vegetables that they are growing to the peak of perfection.

Keep plants healthy Plants that are grown to be strong and vigorous will shrug off mild diseases and pest attacks. Give them first-class soil at the right pH for them. Find them the situation they prefer, in sun or shade. Raise them fast by sowing them at the right time, in their season. Supply them with sufficient water and nutrients but don't overfeed as this will make them soft, sappy and heaven-sent for pests. After a few years, the annual addition of well-rotted bulky organic matter will make the soil so fertile that most plants shouldn't need extra fertilizer. Don't overcrowd them or let them get choked by weeds. Nanny them carefully when they are young and vulnerable, and always avoid stress.

Crop rotation Rotation is vital on allotments where the ground may have been used to grow the same vegetables for decades. Moving crops around to a fresh site each year really helps to prevent colonies of soil-borne pests building up.

Bring in the predators Lure in beneficial predators by providing habitats and food. Ladybirds, hoverflies and lacewings demolish aphids, mites, scale insects, mealy bugs and small caterpillars at an astonishing rate. Ground beetles eat slugs, underground larvae and root aphids. Parasitic wasps eat caterpillars. Centipedes eat slugs and snails. Earwigs eat caterpillars, aphids and codling moth eggs. Frogs, toads, hedgehogs, newts, shrews and slow-worms demolish slugs and many other pests. Birds and bats eat a wide range of pests. Robins love caterpillars, cutworms and other soil-borne grubs. Starlings will go for wireworms and thrushes are skilled at dealing with snails. A well-run garden will achieve a balance.

MAKING HABITATS

Providing water in the shape of a small pond will bring in frogs and toads to breed. It will provide bathing and drinking water for birds and many other creatures. Frogs and toads like to retire for the winter to a secluded spot in the undergrowth. Keep the areas around your plants weed-free but allow a few chosen ones to grow in odd corners as hideaways through winter. Leave some logs and stones in corners for winter refuge. Some allotment sites are fortunate in having hedges, trees and dense evergreen climbers for nesting places around the periphery of the site. Others plant a disused plot with indigenous trees and shrubs as a small nature reserve.

INSECT HOTELS

Parasitic wasps and solitary bees like to hole up in small hollow sticks in winter. It is easy to construct a habitat yourself by drilling out some short lengths of straight sticks and binding them together. Make the bundle watertight and attach it a post, tree or shed in a sunny, sheltered spot. Organic gardening catalogues offer many other insect hotels which are easy to copy.

There has to be an element of live and let live. Birds and wasps are useful in keeping down pests but they will have an eagle eye on your crops.

Grow flowers To lure valuable pollinators and predators into the plot you need to grow flowers. They provide nectar, pollen and habitat for the border patrol of wasps, bees, lacewings and ladybirds. Simple daisy-type flowers in yellow or orange range are particularly favoured. The hoverfly needs an open flower to get to the pollen as it only has a short feeding tube. It lays its eggs on the plant near a colony of aphids and the resulting larvae demolish them.

If you grow buckwheat, clover, lupins or phacelia as green manures, let a few flower as beneficial insects adore them. The flowers of parsnips are equally appreciated if you can wait for them until the following year.

Easy annuals that will attract good predators like a magnet are nasturtiums, marigolds and the poached egg plant (*Limnanthes douglasii*). If you shake the seeds out as the plants go over, they will carry on for years. To extend the season, plant a few wallflowers for spring and Michelmas daisies for autumn.

Camouflage your crops Don't let your plants become a sitting target – disguise them by mixing in flowers and edging your beds with pungent herbs. Many pest insects fly in a random and inaccurate way looking for specific crops. They need a big landing area. If you have small beds with different crops in them they may well miss the target altogether.

Hygiene Diseases can be caused by fungi, viruses, and bacteria. Symptoms include spotting, where parts of the leaves die, cankers or scabs, changing colour (yellowing or silvering), wilting, wet rots and powdery or fluffy moulds or mildew. Viruses are bad news as there is no cure. The symptoms are malformations and unnatural patterns on leaves (mottling or mosaic). They are likely to start on a single plant and be spread by aphids or anything that moves from plant to plant – including you – so good hygiene is essential.

Be careful when accepting gifts of plants. On an allotment it is truly unwise to exchange potatoes, brassicas or onions. They may carry the bad soil-borne diseases eelworm, potato blight, clubroot or onion white rot. Be wary of cross-infection. Dip tools into disinfectant or alcohol when pruning to avoid spreading viruses. Keep pots and containers well scrubbed. Mind where you tread if you have visited another plot where there may be diseases in the soil. Wash your hands between moving from plant to plant.

Nip problems in the bud Many problems can be dealt with if you catch them quickly. Cut away diseased material as soon as it appears before it has a chance to spread. Take care not to shake the spores around. Either burn infected material or put it in a bin liner and get it off the site. Never put it on the compost heap. Pick off pests by hand before they multiply.

Old guttering kept filled with water makes an effective barrier agains slugs and snails

Get to know the pests Some pests are large enough to be seen and identified, many are not or they dine at night. Detection skills come in here, either by night vigils with a torch and magnifying glass or finding clues from the type of damage. If there are holes in the leaves, the chances are that they are being eaten by slugs, snails or caterpillars. If it's the roots, underground larvae might be to blame. If the leaves are curling, most likely there will be colonies of aphids feeding on plant sap on the undersides. If you come in one morning and find your vegetables decapitated it could be the work of cutworms. If some newly planted seedlings have disappeared without trace, suspect mice. Generally speaking, the creatures that move fast are chasing prey and the slow ones are after your vegetables. Not always, however. A ladybird larva, for example, could easily be mistaken for an enemy.

Buy disease-resistant varieties These really help if you have a recurring problem. Where possible use certified virus-free stock.

Outwit pests with barriers and traps Barriers in the form of netting and horticultural fleece are highly effective at keeping out flying insects. Entire crops can be grown under floating mulches, and many experts say that they have transformed their lives. Keep in mind that barriers need to be positioned before the pests arrive and the gauge of mesh needs to be in proportion to the size of the pest. Cloches will keep out slugs as well as flying insects.

Get your timing right Sometimes you can avoid pests and disease by timing alone. Early potatoes are less prone to disease because they grow fast and are out before blight has got going. Over-wintered broad beans are too tough for the black bean aphid to enjoy much, and the first carrots are usually out before the breeding season of the carrot fly.

Use companion planting One area that is waiting for scientific research to prove its worth is companion planting. However, it has been shown that French marigolds

Collars of carpet underlay around individual plants will deter soil-borne creatures, many of which lay their eggs on the soil under the plant for the larvae to feed on when they hatch.

MAKING TRAPS

- ✦ Car grease spread on a board will trap flea beetles when the plant is shaken.
- ✦ Commercially bought sticky grease bands are effective against the wingless winter moths that crawl up apples, pears and plums to lay their eggs.
- ✦ Jars of beer sunk in the ground will send slugs off into drunken oblivion.
- ✦ Pheromone traps lure codling moths looking for a mate.

Members of the onion family and nettles are said to protect their neighbours from diseases caused by fungus and bacteria. Some gardeners make infusions by steeping these plants in water overnight, dousing affected plants and laying the soaked stems around them.

or nasturtiums will mask the smell of brassicas to the cabbage white butterfly, and if four rows of onions are planted around one row of carrots, they keep carrot fly at bay until the onion leaves go over.

Dwarf broad or French beans planted between brassicas of the same size in alternate rows show a lessening in aphid attack. While netting is more full-proof than this, these trials show that combined plantings can have an effect on pests and disease and will help, at the very least, to tip the balance.

Herbs, ancient plants useful to mankind, know how to defend themselves. Tansy contains its own insecticide, and pennyroyal was used in the old days to repel mice and insects in the house. *Tagetes minuta*, a herbal relation of the marigold (also known as Stinking Roger), excretes sulphur compounds from its roots to inhibit eelworms and slugs, as does the mustard green manure *Sinapsis alba*. Lavender and rosemary produce clouds of essential oils that few pests would choose to cross. Chamomile, known as the 'plant's physician', is believed to bring good health to every plant around it.

Biological pest control

Biological controls, or the use of nematodes and parasites to target specific pests, is expensive and temporary. Although effective in an enclosed greenhouse, it is generally impractical for an area as big as an allotment. However, if you have a bad localized infestation, they might be worth considering.

These treatments are usually obtained by post. With one exception (the bacteria *Bacillus thuringiensis* which is sprayed onto caterpillars) they are microscopic forms of life which need to be released by watering onto the soil as soon as possible after delivery. They are highly susceptible to insecticides, so they cannot be used in any joint pest-control programme.

They need moist, well drained soil to do their work and will only work specific temperature bands. Introduce biological controls early so they have time to breed and build up a big enough population to be effective against the pest.

ORGANIC CHEMICALS

Organic chemicals should be used with as much care as their chemical counterparts and only as a last resort. The organic organization, the Henry Doubleday Research Association, only gives the short list below 'qualified acceptance'.

Make sure that you have identified the problem accurately before you use chemicals and target it precisely. The main advantage of organic over inorganic chemicals is that they are non-persistent, in most cases being active for no longer than a day. Don't attempt to make your own chemical treatments as it is illegal in the UK.

GOLDEN RULES FOR USING CHEMICALS

- ✦ Follow the manufacturer's instructions to the letter, particularly when it comes to disposing of chemicals.
- ✦ Store chemicals away from children in the original labelled container.
- ✦ Use good equipment and wash it thoroughly afterwards, having carefully disposed of any leftover chemical.
- ✦ Never spray open flowers for fear of harming bees.
- ✦ Only spray on windless evenings when the good insects have turned in for the night.

Copper fungicide

Copper fungicide acts against the spread of mildews and blights. It coats the leaves for several weeks and comes in various cocktails, including Bordeaux mixture (copper sulphate and quicklime) and Burgundy mixture (copper sulphate with washing soda).

Warning It is toxic to fish and can harm plants, particularly if they are under stress.

Derris

Derris (*rotenone*) comes as a powder or liquid and is made from the derris plant. It is an effective insecticide for aphids, spider mite and other small insects.

Warning It is also poisonous to ladybirds, worms and fish, amongst others.

Insecticidal soap

This is potassium salt soap and is used to control aphids, whitefly, red spider mite and mealy bugs. Only works on a direct hit and is only active for a day.

Warning It will kill friendly insects on contact.

Pyrethrum

Pyrethrum is extracted from *Chrysanthemum coccineum* and is used as an insecticide against aphids, small caterpillars and flea beetle amongst others.

Warning It is poisonous to fish and some friendly insects.

Soft soap

Soft soap is a mild pesticide but is usually used as a medium, or wetter, to help other sprays adhere to leaves.

Sulphur

Sulphur acts as a fungicide. It comes as a spray or dust and can be used to control mildew.

Warning It is harmful to beneficial insects.

COMMON PROBLEMS

Anthracnose Anthracnose is a fungus that can badly affect dwarf and runner beans. The first symptoms are sunken brown spots on the stems. The leaf veins may go red and the leaves drop off. The pods may get infected. Remove and burn or throw away all sick plants, then start again using resistant cultivars.

Aphids There are hundreds of different types of this very common pest which include greenfly and blackfly, mealy cabbage aphid, black bean aphid and lettuce root aphid. They breed like crazy, suck the sap of plants and secrete honeydew. This brings on sooty moulds (a harmless fungus) and they spread viruses as they move from plant to plant.

To control aphids, encourage hoverflies, lacewings and ladybirds which will eat them in huge numbers. Choose resistant varieties. In the case of broad beans, beat the black bean aphid season by planting early under cover. If you do get caught, cut off the tops of the plants where they congregate.

You can also cover plants with horticultural fleece, though not if they need to be pollinated. Squash the aphids or cut off infected shoots and drop into soapy water. Wash them off with a powerful jet of water. Spray bad infestations with soft soap or insecticide soap.

Root aphids are more tricky customers as they work unseen and weaken plants. Different types live amongst the roots of lettuce, runner and French beans and Jerusalem artichokes. They are tiny, yellowish creatures which secrete a fluffy wax. To control root aphids, rotate crops and buy aphid-resistant cultivars.

Asparagus beetle These beetles are chequered black and yellow and the larvae are dark grey. The adults stay hidden around the plants through winter and emerge in spring and lay their eggs. Both adults and larvae will defoliate asparagus and skin the stems. Be watchful from late spring and remove them. Burn the foliage at the end of the season when you cut it down. Clear away hiding places where they may hibernate. As a last resort, use derris liquid or dust.

Bacterial canker *See* canker.

Bacterial leafspot *See* leafspot.

Bean rust Fairly common in warm, damp summers, bean rust is a fungus that causes brownish pustules to grow under the leaves of French and runner beans. If not caught, it will eventually cause the leaves to drop and can spread to the pods. Remove any affected leaves promptly.

Beet leaf miner This shows as large brown patches on the leaves of beetroot and spinach beet. It is caused by the white maggots of the leaf-mining fly. There are two generations each summer. Pick off all the affected leaves or squash the pests within the leaves.

Beet leaf spot This is a mild fungal disease affecting the beet family. Symptoms are small round spots on the foliage, with a purplish edge and pale centre. The over-wintering fungus is spread by rain, tools or hand, and flourishes in hot and humid weather. Clean up the plants, removing affected leaves. Use fresh seed for the next crop.

Birds Pigeons are partial to young brassicas and can desecrate a crop. Use netting and bird scarers. Anything that flaps in the wind will help. Old CDs tied along lines are effective as they swirl and flash as the light catches them. Humming lines also put them off.

Black bean aphid *See* aphids.

Blossom end rot This disease shows as a tough leathery patch at the blossom end, or base, of tomatoes. It may not affect all the fruits on a plant. It is caused by dryness at the roots, or from soil that is too acid. Pick off the affected fruits and water the plants well and regularly.

Bolting When a plant feels under pressure it is inclined to panic and flower before it is ready. Start again with bolt resistant seed, taking care to give it the conditions it needs.

Boron deficiency This is usually found in soils that have been excessively limed, or in thin soils that have become very dry. It can affect root vegetables and also cabbages. In cauliflowers the effect is to make the curds go brown. Improve the moisture-holding capacity of the soil by adding plenty of well-rotted compost ,and keep away from the lime.

Botrytis *See* grey mould.

Cabbage white butterfly The caterpillars of the cabbage white are large hairy yellow creatures with black markings. They feed on the outer leaves of brassicas, leaving holes and stripping a plant with speed.

The larvae of the small white cabbage butterfly are velvety and camouflaged green. They eat the hearts of cabbages. There are two to three generations a year in spring and early autumn.

Watch out for these pests and try to catch them before they burrow in. Pick the affected leaves off. Growing brassicas under fine mesh will help. You can wash them off if you get an infestation with a good dousing of water. The biological control is *Bacillus thuringiensis*.

Cabbage moth These moths only have one generation a year. Their green or light brown caterpillars eat holes in the leaves of brassicas and burrow into the hearts. They also go for onion leaves. Pick them off before they burrow, or grow plants under fine netting. The biological control is *Bacillus thuringiensis*.

Cabbage root fly This is a very serious pest, particularly affecting brassicas, though it can destroy root crops as well, making them inedible. The adults look like small horseflies but it is the offspring of small white maggots that do the damage. The eggs are laid near or on the plants and the pupae overwinter in the soil. Symptoms include wilting and poor growth. The worst damage is likely in late spring but second and third generations can make it a summer-long problem.

Prevention is simple. Put collars of carpet underlay, about 10 cm (4 in) in diameter, around individual plants (see page 179). Alternatively, cover the soil with fleece or fine mesh netting immediately after planting or sowing so the pupae can't burrow.

Cabbage whitefly These sap-feeding insects attack brassicas. The tiny white winged adults fly up when disturbed, while their scaly brown young stay still on the undersides of leaves. A slight infestation causes little harm but don't let it get out of hand.

Symptoms include leaves sticky with honeydew and sooty moulds. The insects are around throughout summer and start laying eggs from mid-May onwards. Clear away all brassicas at the end of the season. Douse infected plants with a good jet of water. It is said that fennel and lovage will attract the parasitic wasp *Aphelinus* which will deal with them effectively. As a last resort, spray weekly with insecticidal soap.

Calcium deficiency This is fairly unusual. It can occur on very acid soils and where there is a lack of water, as plants need moisture to take up calcium. It usually shows as poor growth but can result in blossom end rot.

Carrot fly The adult females lay their eggs around root crops and the resultant larvae tunnel unseen into carrots, celeriac, parsnips, parsley and celery. There are usually two generations, one in late spring and another in midsummer.

The most effective deterrents are to grow the whole crop under fleece or fine mesh, or to erect vertical barriers of fleece, finest mesh, heavy cardboard or sacking at least 60 cm (24 in) high. This will thwart the females as they fly low to the ground and in straight lines.

Sowing in late spring will avoid the first wave and make the second less serious as the population won't have time to build up. Companion planting with onions is said to confuse both the onion fly and the carrot fly which track by scent.

There are disease-resistant cultivars. As the flies don't like the taste but might eat them if there is nothing else, a sacrificial row of ordinary carrots amongst them should guarantee success.

Celery fly These are little white maggots that eat celery leaves. Catch them early and pick off and burn the affected leaves.

Celery leaf spot This disease is caused by a fungus, usually in the seed. It generally appears first as tiny brown spots on the leaves and can spread over the plant rapidly in damp conditions. Cut away any infected parts of the plant and use fresh disease-resistant seed next time.

Chocolate spot This is a fungal disease, affecting broad beans particularly. Small brown spots appear on the leaves and in wet weather it can become serious. Make sure that broad beans have good rich soil, good air circulation and sharp drainage. Pull out any that are badly damaged. Keep the soil clear of weeds and debris. As a last resort, use copper fungicide.

Clubroot A dreadful and widespread fungal disease, particularly affecting brassicas and sometimes root vegetables such as turnips and swedes. There is no cure. The roots become distorted, forming an elbow-like club or a series of tuberous swellings, known as fingers or toes.

If your brassicas are wilting, even though they have enough moisture, dig one up and check the roots. If affected, burn or bin the whole crop without delay and before the disease spreads. Avoid using the same soil for brassicas for as long as possible, keeping in mind that clubroot can live in the soil for 20 years.

Take every precaution to avoid it in the first place. If your soil is acid (the most favourable conditions for clubroot), add lime to bring it up to pH7 or above. Rotation is crucial to avoid a build-up of the disease. Keep weeds under control as some common weeds like shepherd's purse are in the brassica family and can harbour clubroot. Don't use fodder radish and mustard green manures for the same reason.

Be very strict about cleaning tools, and remember that you can spread it on your boots. Buy transplants from a reliable source if you haven't grown them yourself. Grow disease-resistant plants fast, in the right season for them, and the best possible conditions. Start plants in sterile compost, or if sowing outside, make a bigger hole than usual and fill with clean proprietary compost.

Cucumber mosaic virus This disease can affect a wide range of plants, including marrow, courgettes, peppers and spinach, as well as cucumbers. The leaves will develop unnatural yellow patterns and the fruits will be distorted. Burn affected plants, taking great care not to spread the infection. Buy resistant-varieties and remove weeds and debris where the spores will overwinter. Also *see* virus.

Cutworms Cutworms, which can sever the stems of a wide range of young vegetables, are the larvae of a group of nocturnal moths. These moths lay thousands of eggs around the stems of plants in summer. The soil-borne caterpillars, which are brown, yellow or green and curl into a C-shape, will work their way through a row of vegetables by night.

The first signs are plants that have been severed just below the soil level. You will find the caterpillars under the surface of the soil. Clear weeds which the moths like for egg laying and put a collar of carpet underlay around the stems of individual plants. Turn over the soil for the birds to find them.

Downy mildew This disease is caused by a group of fungi which penetrates into plant tissue. They live in the soil for up to five years as well as in plant debris. Discoloured areas appear on leaf surfaces with corresponding white or pale grey fungal growth on the undersides. If allowed to spread, entire leaves will die.

It thrives in damp, warm conditions and is commonly found on a wide range of crops, particularly when young. Remove infected leaves and destroy them. Improve air circulation and clear weeds. Water at the base of plants to avoid spreading it.

Eelworms *See* potato cyst eelworm.

Flea beetles Flea beetles jump. Small and shiny, they emerge in spring and can fly long distances to find food. Young brassicas are their favourite fare and they leave little holes in leaves and stems. Their larvae feed on the roots. A bad attack will check older plants and can kill seedlings.

Protect crops by growing under cloches or fleece. Grow plants as fast as possible in the right season and conditions to get them through the vulnerable period. If you get an infestation, put boards smeared with car grease or slow-drying glue under the plants and shake them. The beetles will leap into the trap. Clear debris round plants. As a last resort, use derris powder.

Grey mould or botrytis This is a common disease prevalent in damp summers, causing leaves, flowers and buds to rot. It's a fluffy mould which, when disturbed, releases clouds of spores, infecting everything around it. Avoid overcrowding, clear any rotting vegetation around plants which will harbour it, and provide a good air flow. Remove all diseased plants and dispose of them.

Halo blight This is a bacterial disease, particularly affecting dwarf French and runner beans. The first symptoms are dark spots in the centre of a pale halo on the leaves. This is followed by yellowing between the veins. The disease comes in the seed and rain spreads it. If you catch it in time, you might save the crop by picking off the leaves. However, next year start again from a fresh source.

Leaf miners The term leaf miner includes many different insects and their larvae that burrow within the leaves of plants. They include the celery and leaf beet miners. You can detect them by the meandering pattern of damage which eventually makes the leaves go dry and brown. If you hold the leaf up to the light you may spot the culprit. Though not generally too harmful to the plant, the thought of eating them is not very appealing. Pick off and destroy the affected leaves.

Leaf mould This disease is usually confined to indoor tomatoes in high humidity. A greenish fungal growth on the undersides of leaves causes yellow patches on the upper sides and, as it spreads up the plant, leaves shrivel and die. It is very infectious. Choose resistant varieties. Destroy diseased material and keep plants well ventilated. Don't water from above. As a last resort, use copper fungicide.

Leaf spot Leaf spot is a fungal disease that affects a wide range of plants, including peas and broad beans. The symptoms are grey or brown spots encircled with rings of discoloured tissue. Sometimes they develop tiny black spots. Leaf spot doesn't do a great deal of damage. Pick off and burn affected and fallen leaves. Make sure the plants have plenty of air circulation and don't water from overhead as this can spread it.

Magnesium deficiency This is caused by acid soil and by overuse of high potash fertilizers. Magnesium is easily washed out by heavy rain. The symptoms are yellowing edges and the areas between the veins of the older leaves. As the leaves deteriorate they may go red, purple, yellow or brown. Foliar feed with Epsom salts, diluting 200 g (8 oz) in 10 litres (2½ gallons) of water with a squeeze of washing-up liquid, fortnightly for immediate effect. Lime the soil to reduce acidity and avoid potash fertilizer if this is the cause.

Mealy cabbage aphids These are sap-feeding insects. The damage shows as yellow patches on the upper surfaces of the leaves. Underneath there will be the pale grey aphids. The effect is to distort the shoot tips and can be very damaging to young plants. All brassicas are vulnerable from spring right through to autumn. Pick the aphids off or give them a good shower of water. If there is an infestation, spray with insecticidal soap.

Mice Mice sometimes steal newly planted seedlings without trace and raid your shed. Use cloches to protect plants or trap the mice.

Mildews *See* downy mildew and powdery mildew.

Mosaic virus *See* virus and cucumber mosaic virus.

Onion fly Having spent the winter below ground, onion flies emerge as adults in May. They lay their eggs on and around the host plants – onions in particular, but leeks, shallots and garlic also. The resulting small white maggots cause havoc to the crops. They bore holes into them, eat the roots and cause the plants to rot. Remove any affected plants and destroy them. Outwit the flies by rotation, plant onion sets rather than seedlings, and cover plants with fine net or fleece before the flies arrive.

Onion neck rot This is a fungal disease exacerbated by wet weather. The signs of damage don't usually appear until the onions have been in store for a couple of months, when they become soft and discoloured. A grey, fluffy, fungal growth develops, particularly around the neck of the bulb.

The fungus lurks in onion debris and on the soil. Grow your onions to the book. Use top-quality sets, and allow for plenty of air circulation while they are growing and during storage. Don't store any that are damaged, and practice rotation.

Onion white rot This is a fatal disease caused by a fungus in the soil. The whole onion family is susceptible. A fluffy white fungus dotted with little black specks spreads over the base of the bulbs and the roots and the leaves turn yellow and wilt. The black specks are 'sklerotia', or 'fruiting bodies', which will lurk in the soil for up to seven years waiting for the next host. Dispose of the crop and avoid using the same land for the onion family for a good eight years.

Parsnip canker Parsnip canker shows as rough brown, red or black patches usually on the shoulders of the roots. It is a fungal infection which enters through a wound, perhaps a tiny hole left by a carrot fly. There isn't a cure so it is a good idea to buy disease-resistant seed.

Pea and bean weevil If you find U-shaped holes in the edges of your pea and bean leaves, it will be the tiny, grey pea and bean weevil. The larvae live in the soil and feed on the roots, and adults emerge in June or July. Generally, they don't do too much harm except to young and vulnerable plants. The best defence is to cover with cloches or net when you plant.

Pea moths The larvae of pea moths are tiny caterpillars with black heads. The adults lay eggs on the flowers to hatch out inside the pea pod. Avoid damage by planting peas early or late to avoid the breeding time. Cover plants with fine mesh when in flower. Turn over the soil for the birds to get at them in winter. As a last resort, use derris liquid a week after flowering and again two weeks later.

Potato blackleg This disease is caused by soil-borne bacteria. It first shows as stunted leaves and blackened rot at the base of the stem. The parent potato will also be rotting. It enters the potato through a break in the skin and may be isolated to a single potato plant amongst a crop. It's more likely in wet conditions. Avoid damaging potatoes, particularly when lifting them. Only store tubers with unbroken skin and buy certified seed potatoes.

Potato blight *Phyptophtora infestans* is the fungus that caused the Irish potato famine in 1846. It affects potatoes and tomatoes and shows as brown marks on leaves and stems and a downy white mass of fungal spores on the undersides. These wash down into the soil and the tomato fruits and potato tubers develop sunken areas and tomatoes can get leathery patches. The foliage blackens and rots. Other fungi and bacteria join in to induce a fast-spreading highly infectious soft rot. It is more common in wet summers and generally doesn't affect early crops.

Remove and destroy any infected leaves. Buy resistant strains and good quality seed potatoes. Space plants widely for air circulation and water at ground level. Earth up and mulch potatoes with straw to prevent the spores reaching the tubers. Dig up all potatoes at the end of the season. Burn or destroy affected plants as any remaining debris will infect the next crop.

Potato common scab This is a bacterial disease usually occurring in dry weather and on sandy soils, particularly if newly cultivated. Circular scabby patches appear on the tubers, which are harmless if disfiguring and can usually be peeled off. Increase the acidity of the soil and the water-holding capacity with well-rotted organic matter. Keep the crop well watered. Buy resistant cultivars.

Potato cyst eelworm This pest is to be avoided at all costs as there is no cure. Each microscopic nematode female can lay 500 eggs, which can stay in the ground for ten years or longer waiting for a host plant to arrive before they hatch. On an allotment where much potato-growing has taken place over years, eelworm eggs are likely to be lurking in the soil.

The larvae feed on the roots. The first symptoms are drying, dying leaves starting from the bottom, followed by poor crops. If allowed to build up, there will be crop failure.

Look for varieties with resistance where possible. Avoid bringing the eelworm in on your boots or tools from other parts of the allotment site that may be infected. Keep up the rotation with as big a gap as possible before returning to the original potato patch. Put on lots of well-rotted compost and manure to bring in nematode predators. When the ground is fallow, plant a mustard green manure (*Sinapsis alba*) as it inhibits eelworms.

Potato powdery scab This fungal disease is mostly found on heavy soils where potato crops have been grown over the years. The potatoes form scabby patches, which burst to release thousands of spores into the soil. There is nothing for it but to destroy the crop and abandon the site for potato-growing for three or four years, only returning to it when the drainage has been improved.

Powdery mildew Powdery mildew shows as a whitish powdery growth – typically on the upper sides of the leaves but can be almost anywhere – on a wide variety of plants. If left to develop it can cause yellowing, distortion, dead spots on leaves and leaf drop. In severe cases it can kill. It thrives particularly on young plants in dry soil in humid conditions.

Catch it early and remove affected leaves. Keep the plants watered but avoid spreading the fungi by watering from overhead. Don't overfeed plants with high nitrogen fertilizers as they will produce soft lush growth. Look out for cultivars with disease resistance. As a last resort, spray with copper fungicide – Bordeaux mixture or Burgundy mixture.

Red spider mite This seems to be increasing in our warmer summers. They are microscopic, brown mites flaring to brilliant orange-red in winter. Traditionally associated with the greenhouse, they sometimes get into the garden in summer. The outdoor sap-feeding fruit tree version is usually kept under control by natural predators. Their presence is revealed by a silvering of the leaves, followed by a mottled yellowing. They like dry conditions best, so a good squirt of water on a regular basis will help to keep them at bay with or without the addition of insecticidal soap.

Root aphids *See* aphids.

Scab *See* potato powdery scab and potato common scab.

Shothole This is a general term for bacterial and fungal infections that cause holes edged with a brown ring to appear in leaves. Remove affected leaves and destroy. Keep up air circulation and remove weeds.

Slugs Slugs are a universal menace. There are many different species; some live underground and attack roots and burrow into potato tubers. Slugs like warm moist weather and come out to feed at night. They shelter and breed under stones, in piles of leaves, under garden debris, in the soil, and under mulches – the one disadvantage.

Encourage predators including frogs, toads, birds, hedgehogs, shrews, slow worms and beetles. Cover small plants with cloches or surround plants with inhospitable mulches, such as sharp gravel and egg shells, or dry woodchips, soot, ash, or lime. Do torchlight vigils and catch and destroy slugs by dropping them into a salt solution. Using tongs or rubber gloves makes this slightly less repellent.

Trap and drown them by sinking plastic pots of beer or milk into the ground, changing the bait from time to time and stopping if you catch beetles. Slugs will gorge until they die on bran.

Scooped-out melon and grapefruit skins make hiding places for them to congregate as do old planks or wet newspaper. Lay out alternative food sources. They will choose rotting vegetation over fresh, so old lettuce leaves kept moist under cover of a tile will draw them away to where they can be collected.

Snails Snails present the same problems as slugs and the same treatments are effective, though frogs won't eat them.

Soft rot This bacterial and fungal disease results in decay of plant tissues. It affects swedes and turnips, showing as a greyish mushy rot on heavy wet ground. Start again, trying raised beds for better drainage.

Sooty mould Sooty mould results from the honeydew excreted by aphids, whiteflies and mealy bugs. Though unsightly (the leaves get covered with grey to black mould), it is not in itself harmful except that it reduces light and air getting to the leaves. The only way to deal with it is to go after the pests.

Tip burn This can affect lettuce, chicory and cauliflowers, making the leaf edges turn brown. It is usually caused by calcium deficiency due to dry roots or the soil being too acid. Water well and if your soil is acid, lime it before planting lettuce next time.

Tomato blight *See* potato blight.

Violet root rot This shows as filaments of violet fungal threads covering the roots, crown and stems of plants. It can affect asparagus, beetroot, carrots, celery, parsnips, potatoes, strawberries, swedes and turnips. The first signs are yellowing and stunting. It's most commonly found in wet, acid soils. Remove and destroy the crop and change the soil conditions before trying again.

Vine weevils Vine weevil larvae live in the soil. The white grubs feed on plant roots, causing poor growth and total collapse in bad cases. The black adults are less harmful – appearing at night to nibble the outside edges of the leaves, they leave small ragged holes. Most die by winter. Predominantly female, they lay hundreds of eggs in the soil through the summer months. These pests have become resistant to destruction.

Viruses Viruses come in many forms, all incurable. The first signs are loss of vigour, stunting and distortion, followed by strange colour changes and patterns on leaves including mosaic patterns, flecking and mottling. Viruses can be caused by soil-borne nematodes and fungi. More commonly, they will be passed from plant to plant by aphids or even gardeners as they move around handling different plants. Try to prevent the spread by strict hygiene and keeping the aphid population at bay. Buy certified virus-free stock where possible. Destroy and burn infected plants as soon as virus is detected.

(Also *see* cucumber mosaic virus).

Whiptail Whiptail causes leaves to become mottled and yellow, particularly in brassicas. It is caused by a lack of the trace element molybdenum. It is more likely to arise in acid soils. The solution is to lime the soil.

Whitefly These are tiny flying insects which suck the sap on the underside of leaves and excrete copious amounts of sticky honeydew. If you shake the plants you can suck them up with portable vacuum cleaner or spray them with insecticidal soap.

Wireworms Wireworms are orange click beetle larvae that live underground and bore holes into carrots, brassicas, lettuces, onions, tomatoes and potatoes. They are mostly found in grassland and the good news is that, as your allotment becomes more cultivated with less grass, they will diminish in number. You can trap them by burying carrots, cabbage or a potato on a stick to attract them away from your crops. Keep the ground weed-free and dig it over in winter for the birds to find them.

USEFUL ADDRESSES

Seed suppliers

Chase Organics
Tel order line: 0845 130 1304
www.OrganicCatalogue.com.
The official mail order catalogue for the Henry Doubleday Research Association. Has a comprehensive selection of organic seed.

Chiltern Seeds
Tel: 01229 581137.
www.chilternseeds.co.uk.
Good list of vegetables, including orientals and interesting varieties.

Dobies
Tel: 0870 1123625
www.dobies.co.uk
A good selection, including 'pot ready plants'.

Edwin Tucker & Sons
Tel: 01364 652233.
www.edwinbtucker.com
Selection of old and new varieties. Discounts for associations.

Kings Seeds
Tel: 01376 570000
www.kingsseeds.com
Official supplier for the National Society of Allotment and Leisure Gardeners.

Mr Fothergill's Seeds
Tel: 0845 1662511
www.mr-fothergills.co.uk
Selection of heritage, organic, oriental. Club discounts

Seeds by Size
www.seeds-by-size.co.uk
8,000 varieties of vegetable, herb and flower seed.

Suffolk Herbs
Tel: 01376 572456
www.suffolkherbs.com
Good list of organic vegetable seed including unusual varieties and ancient grain crops.

Suttons
Tel order line 0870 220 2899
www.suttons-seeds.co.uk
Established 1806 and now the biggest seed company in the world.

Tamar Organics
Tel: 01822 834690
Specializes in organic products.

Terre de Semences with Association Kokopelli
Tel: 01227 731815
www.terredesemences.com
Offers a wide ranging selection of interesting organic seed.

Thomas Etty Esq.
Tel: 020 8466 6785
www.dircon.co.uk
Small firm, specializing in heritage, unusual and regional vegetable seed.

Thompson & Morgan
Tel: 01473 688821
www.thompson-morgan.com
Good choice in vegetables. Specializes in 'plugs'.

Unwins
Tel: 01945 588522
Small but interesting collection of old and new vegetables.

Specialist suppliers for the showbench

Select Seeds
Tel: 01246 826011
www.selectseeds.co.uk

Shelley Seeds
Tel: 01244 317165

Giant vegetable seed

W. Robinson & Sons
Tel: 01524 791210
www.mammothonion.co.uk

National societies and organizations

Henry Doubleday Research Association (Garden Organics)
Tel: 024 7630 3517
The organic organization. Open gardens, magazine, events, opportunity to join the Heritage Seed Library. Excellent advice line for members.

National Institue of Agricultural Botany (NIAB)
Tel: 01223 342200
www.niab.com
An independent bio-science company, involved with the development and use of plant genetic resources in the agricultural and food industries. It produces a very useful booklet for the amateur entitled the *NIAB Veg Finder.* Aimed to simplify the hunt for particular varieties, it lists more than 5,000 vegetables and herbs.

National Pot Leek Society
Tel: 0191 549 4274

National Society of Allotment and Leisure Gardeners
Tel: 01536 266 576

National Vegetable Society
Tel: 0161 442 7190
www.nvsuk.org,uk

Royal Horticultural Society
Tel: 020 7834 4333
www.rhs.org.uk

INDEX (page numbers in *italic* refer to the illustrations)